my **revision** notes

WJEC EDUQAS GCSE (9–1)

DESIGN AND TECHNOLOGY

Ian Fawcett
Jacqui Howells
Andy Knight
Chris Walker

HODDER
EDUCATION
AN HACHETTE UK COMPANY

Orders: please contact Hachette UK Distribution, Hely Hutchinson Centre, Milton Road, Didcot, Oxfordshire, OX11 7HH. Telephone: +44 (0)1235 827827. Email education@hachette.co.uk Lines are open from 9 a.m. to 5 p.m., Monday to Friday. You can also order through our website: www.hoddereducation.co.uk

ISBN: 978 1 5104 7169 6

First published in 2019 by
Hodder Education,
An Hachette UK Company
Carmelite House
50 Victoria Embankment
London EC4Y 0DZ
www.hoddereducation.co.uk

Impression number   10   9   8

Year   2023

Cover photo © nikkytok – stock.adobe.com

Typeset in India.

Printed in Spain.

A catalogue record for this title is available from the British Library.

MIX
Paper | Supporting
responsible forestry
FSC™ C104740

Get the most from this book

Everyone has to decide his or her own revision strategy, but it is essential to review your work, learn it and test your understanding. These Revision Notes will help you to do that in a planned way, topic by topic. Use this book as the cornerstone of your revision and don't hesitate to write in it: personalise your notes and check your progress by ticking off each section as you revise.

Tick to track your progress

Use the revision planner on page iv to plan your revision, topic by topic. Tick each box when you have:

- revised and understood a topic
- tested yourself
- practised the exam questions and checked your answers.

You can also keep track of your revision by ticking off each topic heading in the book. You may find it helpful to add your own notes as you work through each topic.

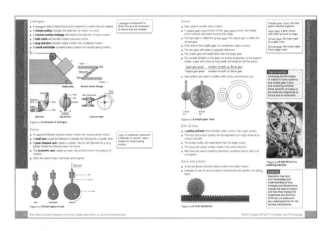

Features to help you succeed

Exam tips

Expert tips are given throughout the book to help you polish your exam technique in order to maximise your chances in the exam.

Typical mistakes

The author identifies the typical mistakes candidates make and explains how you can avoid them.

Key words

Clear, concise definitions of essential key terms are provided where they first appear, and in the Glossary at the back of the book.

Now test yourself

These short, knowledge-based questions provide the first step in testing your learning. Answers are available at www.hoddereducation.co.uk/myrevisionnotes

Exam practice

Practice exam questions are provided for each topic. Use them to consolidate your revision and practise your exam skills.

My revision planner

Acknowledgements

Picture credits

The Publishers would like to thank the following for permission to reproduce copyright material.

Fig.1.1 © Stepan Popov/123 RF; Fig.1.3 © Ulldellebre/stock.adobe.com; Fig.1.4 © European Union (see https://eur-lex.europa.eu/legal-content/EN/TXT/?uri=CELEX:32010R0066 and https://ec.europa.eu/environment/ecolabel/index_en.htm); Table 1.3 *t* © Petovarga/123 RF *b* © Petovarga/123 RF; Fig.1.5 © Plus69/stock.adobe.com; Fig.2.2 © Alexandr Bognat/stock.adobe.com; Fig.2.3 © RichLegg/E+/Getty Images; Fig.2.4 © Colin Moore/123 RF; Fig.4.5 © Nikkytok/stock.adobe.com; Fig.4.6 © Vladimir/stock.adobe.com; Fig.5.4 © Digital Genetics/Shutterstock.com; Fig.5.6 © Ruzanna Arutyunyan/stock.adobe.com; Fig.6.13 © Andrey Popov/stock.adobe.com; Fig.6.15 © Varin Rattanaburi/123 RF; Fig.6.17 © Danielle Nichol/Alamy Stock Photo; Fig.7.4 © Kitch Bain/Shuttetstock.com; Fig.8.4 © Justin Kase z12z/Alamy Stock Photo; Fig.8.8 Ian Fawcett; Fig.8.9 Ian Fawcett; Fig.8.10 Ian Fawcett; Fig.9.4 Ian Fawcett; Fig.9.5 Ian Fawcett; Fig.9.6 © Treboreckscher/123 RF; Fig.9.8 © Alexlmx/stock.adobe.com; Fig.9.11 © Hoda Bogdan/stock.adobe.com; Fig.9.18 © Mbongo/stock.adobe.com; Fig. 9.19 © Gl0ck33/123 RF; Fig. 9.20 © Africa Studio/stock.adobe.com; Fig.10.3 © Pressmaster/stock.adobe.com; Fig.10.4 © Unkas Photo/Shutterstock.com; Fig.10.5 © Anton Starikov/Shutterstock.com; Fig.10.6 © C R CLARKE & CO (UK) LIMITED; Fig.10.7 © Shaffandi/123 RF; Fig.10.8 © Serhii Barylo/stock.adobe.com; Fig.10.9 © Bravissimos/stock.adobe.com; Fig.10.10 © Sorapolujjin/stock.adobe.com; Fig.10.14 © Ajay Shrivastava/stock.adobe.com; Fig.10.15 Ian Fawcett; Fig.10.16 Ian Fawcett; Fig.11.5 © Chamillew/stock.adobe.com; Fig.11.8 © Sergojpg/stock.adobe.com; Fig.11.12 © Nataliia Pyzhova/stock.adobe.com; Fig.11.13 Jacqui Howells; Fig.11.14 © Pincasso/stock.adobe.com; Fig.11.15 Jacqui Howells; Fig.11.16 © Tapui/stock.adobe.com; Fig.11.17 Jacqui Howells; Fig.11.19 © Vvoe/stock.adobe.com; p.121 © Bildlove/stock.adobe.com; Fig.12.1 © Wolf Safety Lamp Company; Fig.13.1 © AntonioDiaz/stock.adobe.com; Fig.15.1 © Pixtour/stock.adobe.com; Fig.17.1 © Steve Mann/123RF; Fig.17.2 © Dyson; Fig.17.3 © Stefan Rousseau/PA Archive/PA Images; Fig.19.7 Dan Hughes; Fig.20.1 Andy Knight; p.142 KITEMARK and the Kitemark device are reproduced with kind permission of The British Standards Institution. They are registered trademarks in the United Kingdom and in certain other countries; Fig.22.1a © Peter Polak/stock.adobe.com; Fig.22.1b © Bookzaa/stock.adobe.com; Fig.22.2 Chris Walker; Fig.22.3 © Arvind Balaraman/123 RF; Fig.22.4 © Michał Dzierżyński/123 RF; Fig.22.5 © Nndrln/stock.adobe.com; Fig.22.6 © Mny-Jhee/stock.adobe.com

Countdown to my exams

6–8 weeks to go

- Start by looking at the specification — make sure you know exactly what material you need to revise and the style of the examination. Use the revision planner on pages iv and v to familiarise yourself with the topics.
- Organise your notes, making sure you have covered everything on the specification. The revision planner will help you to group your notes into topics.
- Work out a realistic revision plan that will allow you time for relaxation. Set aside days and times for all the subjects that you need to study, and stick to your timetable.
- Set yourself sensible targets. Break your revision down into focused sessions of around 40 minutes, divided by breaks. These Revision Notes organise the basic facts into short, memorable sections to make revising easier.

REVISED ☐

2–6 weeks to go

- Read through the relevant sections of this book and refer to the exam tips, exam summaries, typical mistakes and key terms. Tick off the topics as you feel confident about them. Highlight those topics you find difficult and look at them again in detail.
- Test your understanding of each topic by working through the 'Now test yourself' questions in the book. Look up the answers at **www.hoddereducation.co.uk/myrevisionnotes**.
- Make a note of any problem areas as you revise, and ask your teacher to go over these in class.
- Look at past papers. They are one of the best ways to revise and practise your exam skills. Write or prepare planned answers to the exam practice questions provided in this book. Check your answers online and try out the extra quick quizzes at **www.hoddereducation. co.uk/myrevisionnotes**
- Use the revision activities to try out different revision methods. For example, you can make notes using mind maps, spider diagrams or flash cards.
- Track your progress using the revision planner and give yourself a reward when you have achieved your target.

REVISED ☐

One week to go

- Try to fit in at least one more timed practice of an entire past paper and seek feedback from your teacher, comparing your work closely with the mark scheme.
- Check the revision planner to make sure you haven't missed out any topics. Brush up on any areas of difficulty by talking them over with a friend or getting help from your teacher.
- Attend any revision classes put on by your teacher. Remember, he or she is an expert at preparing people for examinations.

REVISED ☐

The day before the examination

- Flick through these Revision Notes for useful reminders, for example the exam tips, exam summaries, typical mistakes and key terms.
- Check the time and place of your examination.
- Make sure you have everything you need — extra pens and pencils, tissues, a watch, bottled water, sweets.
- Allow some time to relax and have an early night to ensure you are fresh and alert for the examination.

REVISED ☐

My exam

Component 1: Design and Technology in the 21st Century

Date:..

Time:..

Location:...

1 Technical principles: core knowledge and understanding

1 Design and technology and our world

The impact of new and emerging technologies on industry and enterprise

REVISED

- Following the industrial revolution, the use of steam to provide power led to the development of new machinery and manufacturing equipment which made products more quickly and efficiently.
- The use of electricity to power machinery led to products being **mass produced** on **assembly** lines.
- Modern factories increasingly use **automated production**. Robots are used to complete repetitive and monotonous tasks instead of humans. Productivity and product quality are improved as a result of automation.

> **Mass produced**: hundreds or thousands of identical products manufactured on a production line.
>
> **Assembly lines**: a line of equipment/machinery manned by workers who gradually assemble a product as it passes along the line.
>
> **Automated production**: the use of computer-controlled equipment or machinery in manufacturing.

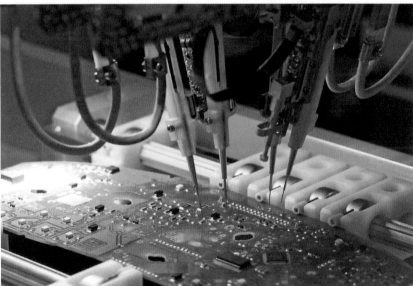

Figure 1.1 Robots are used in the production of printed circuit boards

Market pull and technology push

- Some products are developed from market research, which uncovers a need for a new product or a demand for an improved existing product. This is referred to as **market pull**.
- Technological developments in materials, components or manufacturing methods leads to the development of new or improved products. This is called **technology push**. For example:
  - many touch screen devices only exist because of the development of the material graphene
  - the technology now exists to weave **conductive** threads into fabric for clothing, which when worn will interact directly with the wearer.

> **Market pull**: a new product is developed in response to a demand in the market or users.
>
> **Technology push**: products developed as a result of new technology.
>
> **Conductive**: the ability to transmit heat or electricity.

Consumer choice

- Designers and manufacturers respond to customer choice by developing products that specifically meet the needs of consumers.
- Many people will only want to buy products that include the latest technology, for example, mobile phones are replaced because products with the latest technology become available.

Product life cycle

A product's **life cycle** is a marketing strategy that looks at the four main stages a product goes though.

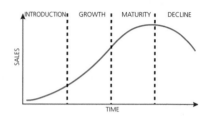

Figure 1.2 The product life cycle

Table 1.1 The four main stages of a product's life cycle

Introduction	Following an advertising campaign, the new product is introduced to the marketplace
Growth	Sales will grow as consumers become aware of the product and buy it
Maturity	Sales are at their peak with companies hoping to achieve maximum sales for the product
Decline	Sales begin to fall as most interested consumers now own the product or a new product has replaced it

> **Life cycle**: the stages a product goes through from beginning (extraction of raw materials) to end (disposal).
>
> **Fad product**: a product that is highly popular for only a very limited amount of time.

- The length of a product life cycle will depend on the product:
  - classic fashion styles maintain good sales for many years, while fashion **fad products** become popular very quickly but decline just as quickly as new styles are introduced
  - examples of recent fad products include loom bands and fidget spinners.

People, culture and society

Global production and its effects on people and culture

- Developments in transport makes it easier for manufacturers to ship materials, components and products all over the world.
- The global economy allows for materials and components to be sourced in one country, manufactured into products or part-products in another and then shipped all over the world.
- Manufacturing costs can be reduced through automation.
- Developments in mobile technology and the internet make it easier to communicate with people all over the world. This leads to greater competition among manufacturers, which in turn keeps prices low.
- There are downsides to this global society which directly affect workers:
  - importing cheap products from overseas and not buying locally-produced products can lead to job losses in our own society
  - the use of mobile technology can make people feel isolated as the opportunities for face-to-face interaction are limited
  - the use of automation in manufacturing leads to job losses. Fewer people are needed in factories
  - workers overseas are often paid low wages in an effort to keep costs down and maximise profit for the manufacturer.

- Designers need to be aware and sensitive to cultural differences. What is acceptable in one country may be offensive in another. The values, beliefs and customs of different cultures should be respected.
- Global production is also a threat to traditional industries, skills and techniques in some developing countries.

Legislation and consumer rights

- The Consumer Rights Act 2015 protects consumers when goods purchased or services provided are not as expected.
- The Act also covers digital products and buying online, and covers contracts such as those issued with mobile phones.
- The law states that all goods should be as described or seen when purchased and be fit for purpose.
- The Act protects consumers against faulty or **counterfeit** goods, and poor service or problems with builders; this includes rogue traders.
- Consumers can request a refund, repair or replacement when goods purchased do not meet certain standards:
  - ○ the product should function as intended
  - ○ the product should be of satisfactory quality
  - ○ the product should be as described at the time of purchase.
- If a service provided fails to meet expectation and a full refund or replacement is not possible, then the provider is legally bound to offer some form of **compensation**.

> **Counterfeit**: an imitation of something, sold with the intent to defraud.
>
> **Compensation**: payment given to someone as a result of loss.

Moral and ethical factors

- A global market allows unrestricted trade. Many people grow their businesses, make a profit and improve the lives of their workers by offering regular employment and an income.
- There is no obligation on companies to improve the lives of their workers. Some companies put profit above all else, with low wages for workers and poor working conditions.
- Some companies follow a more ethical approach to trade. They focus on goods and services that are beneficial to consumers, they show that they are socially responsible by treating their workers fairly with acceptable rates of pay and good working conditions, and they support environmental causes.
- Ethical traders are open and transparent about costs. It is important to them that trade is seen to be fair.
- Some companies choose not to disclose their costs because this could reveal poor wages or working conditions if profits are revealed and considered high. This is particularly true in the textile industry.

Sustainability and the environment

- Designers, manufacturers and consumers are increasingly aware of the negative impact new technologies and the development and disposal of products has on the environment.
- **Sustainability** is about meeting today's needs without compromising the needs of future generations.
- It is important to look at ways of reducing our environmental impact.

> **Sustainability**: meeting today's needs without compromising the needs of future generations.

- Developments in renewable energy allow us to reduce our reliance on **finite fossil fuels** such as coal or oil.
- Many plastics are difficult to recycle. New technology is being developed that will allow these plastics to be broken down more effectively and safely, allowing them to be recycled.

> **Finite fossil fuels**: a limited amount of natural resources that cannot be replaced.

Production techniques and systems

Computer-aided design (CAD)

- All aspects of developing design ideas through to 3D models can be done on the computer.
- CAD packages allow for changes to be made or errors to be rectified.
- CAD models allow designers and manufacturers to simulate how products will look and perform in different situations.
- Emerging **cloud-based technology** allows for collaborative work – designers can share projects via the internet. This collaboration can be on a global scale, which cuts the need for travel.
- **Generative design** is a design development that makes use of mathematical **algorithms** based on set parameters or design requirements.
- Disadvantages of CAD include high initial set-up costs, including training employees, and the possibility of losing work through computer failure or virus.

> **Cloud-based technology**: technology that allows designers to share content via the internet.
>
> **Generative design**: a computer-based iterative design process that generates a number of possibilities that meet certain constraints, including potential designs that would not previously have been thought of.
>
> **Algorithms**: a logical computer-based procedure for solving a problem.

Computer-aided manufacture (CAM)

- CAM machinery can manufacture products and components directly from CAD drawings.
- CAM machines are often used where high volumes of identical products of a consistent high quality are needed.
- Initial set-up costs can be high, but they are considered more efficient in the long term as they can run for long periods without breaks.
- Disadvantages of CAM include high set-up costs, impact on workforce with loss of employment, technological failure and on-going maintenance costs.

> **Additive manufacture**: computer-controlled manufacture of a 3D object by adding materials together layer by layer.

Table 1.2 Types of CAM equipment

CNC embroidery machine	Designs can be embroidered directly onto a range of textile fabrics Designs can be saved then repeated several times with the same high-quality finish
Vinyl cutters	A pattern based on a CAD drawing can be cut from a roll of self-adhesive vinyl Letters used in signage are often cut on a vinyl cutter; the colour is determined by the vinyl
CNC router	A rotating router cutter follows a CAD drawing to cut a path or shape Alternative cutting tools can alter the profile of the cut The CAD drawing dictates the depth of the cut
Laser cutter	Laser cutters use a laser beam to cut through (or vaporise) material; material can also be engraved Intricate patterns can be cut from a variety of materials. However, not all materials can be cut as some such as nylon and PVC can burn or melt
3D printer	3D printing (also known as **additive manufacture**) uses a thermoforming polymer roll or spool of filament which is heated, then extruded through a head to form a layer. The bed then moves down for the next layer to be printed The strength of the product is determined by the inner design of the print and the material used

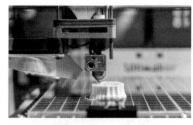

Figure 1.3 3D printing or additive manufacture

How the critical evaluation of new and emerging technologies informs design decisions

REVISED

Sustainability and environmental issues

- Designers must take the environment into consideration when making decisions. For example:
  - choosing materials that are more environmentally friendly
  - manufacturing products using efficient, low-energy processes
  - ensuring better build quality so that products last longer
  - using less packaging, or using recycled packaging
  - reducing transportation by using local manufacturers with locally-sourced materials
  - using LED lights instead of filament lamps
  - avoiding products with a short life cycle
  - making recycling of products easier
  - considering Fairtrade products where everyone in the supply chain is treated fairly.
- **Environmental directives** (laws), which come from the European Union or organisations such as the World Energy Council, are targets for all countries to work towards to reduce energy consumption, reduce pollution and eliminate the disposal of hazardous waste into the environment. These directives also cover climate change, air pollution and the protection of wildlife.

Figure 1.4 Products, goods and services awarded the EU Ecolabel have met a tough set of environmental standards

> **Environmental directive**: a law to provide protection for the environment.

Social, cultural, economic and environmental responsibilities

- Designers and manufacturers need to consider the views of consumers; demand for more environmentally-friendly products is increasing.
- In an effort to reduce energy consumption, domestic appliances carry an energy rating: A+++ is the most efficient while G is the least. Using efficient products will also help reduce household energy bills.

Linear and circular economy

Table 1.3 Linear and circular economy

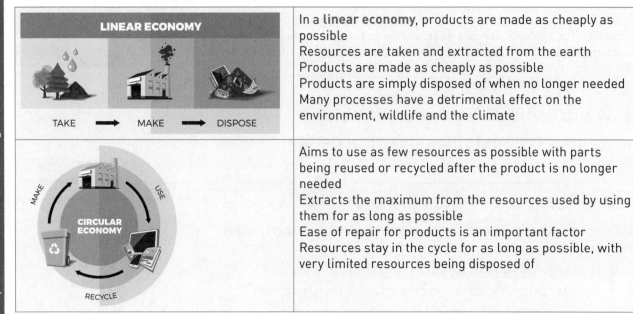

	In a **linear economy**, products are made as cheaply as possible Resources are taken and extracted from the earth Products are made as cheaply as possible Products are simply disposed of when no longer needed Many processes have a detrimental effect on the environment, wildlife and the climate
	Aims to use as few resources as possible with parts being reused or recycled after the product is no longer needed Extracts the maximum from the resources used by using them for as long as possible Ease of repair for products is an important factor Resources stay in the cycle for as long as possible, with very limited resources being disposed of

- Manufacturers are having to reconsider the way they source, manufacture and package products.
- Laws are in place to tackle pollution and how waste is disposed of.
- **Cradle-to-cradle production** links with the **circular economy** – from the source of materials to a product's rebirth as a new product there is little or no waste.
- **Cradle-to-grave production** includes consideration of how the product will ultimately be disposed of.

The six Rs of sustainability

The type of questions a designer might ask that could reduce the environmental impact of a product are outlined in Table 1.4.

Table 1.4 The six Rs of sustainability

Rethink	Is there a better way of making the product that is less harmful to the environment? Can the design be simplified to make manufacturing easier?
Recycle	Can the product be recycled easily after it is no longer needed? Can the components and materials be separated easily? Can recycled materials be used?
Repair	Can this product be repaired easily if it breaks? Can the component parts be replaced easily?
Refuse	Consumers might not buy the product if it is not environmentally friendly. Where will it be made and what are the conditions for workers? Is the product unethical?
Reduce	Can the number of component parts or new materials be reduced? Can packaging be reduced? Can the manufacturing process be simplified to reduce the energy used?
Reuse	Can any part be reused after it is no longer needed? Could the product be reused for something else once its primary use is no longer required?

Linear economy: raw materials are used to make a product and waste is thrown away.

Circular economy: a system that aims to minimise waste and extract the maximum possible use from resources.

Cradle-to-cradle production: considering the product's complete life cycle, including the reuse or recycling of materials after its initial use has ended.

Cradle-to-grave production: considering the product's complete life cycle until it is disposed of.

Now test yourself answers at www.hoddereducation.co.uk/myrevisionnotes

Life-cycle analysis (LCA)

- A **life-cycle analysis** looks at the environmental impact of a product throughout its entire life: from the source of materials, through its useful life to final disposal and potential rebirth as a new product.
- The following factors should be considered: source of raw materials, material processing, manufacturing, use, end of life and transportation, including energy used at various points during the product's life cycle.

Fairtrade

- Fairtrade sets up partnership schemes between producers, businesses and consumers.
- Social, economic and environmental standards are set for all companies, producers and workers involved in the supply chain.
- By giving workers a share of the profits or fairer wages, the lives of workers will be improved and it will help combat poverty.
- Conditions for workers must be of a satisfactory standard, no one is **exploited** and workers' rights are protected.
- Consumers can be satisfied that buying products that carry the Fairtrade Mark supports disadvantaged workers and producers in developing countries.

Carbon footprint

- **Carbon footprint** is a measure of the total amount of greenhouse gases produced as a result of human activity, which includes the manufacturing of products.
- Greenhouse gases are usually measured in units of carbon dioxide (CO_2) and are said to be the cause of **global warming**.
- When we use energy that comes from fossil fuels, we add to our carbon footprint – for example, heating our homes or workplaces with gas, oil or coal, transporting products or travel by car or aeroplane.
- Designers can reduce the carbon footprint of a product by adopting more sustainable approaches to design, for example, sourcing locally produced materials to cut down the transportation of raw materials.

Ecological footprint

- Ecological footprint is a measure of the impact human activity has on the environment.
- We depend on the world's natural resources and productive land to enable us to produce goods and services.
- The products and clothes we use, the food we eat, the waste we generate and the way we live all contribute to our ecological footprint.
- Biologically productive land is continually being cleared to sustain our lifestyles and an increasing global population.
- Humanity's ecological footprint is currently 1.7 Earths.
- If we continue to use up the world's natural resources quicker than nature can replace them, we will create an **ecological deficit**.

> **Typical mistake**
>
> Do not confuse life-cycle analysis with a product life cycle. Life-cycle analysis looks at the environmental impact of a product over its entire lifetime whereas the product life cycle is a marketing strategy looking at sales of products.

> **Life-cycle analysis**: an assessment of a product's environmental impact during its entire life.
>
> **Exploit**: unfairly take advantage of.

> **Carbon footprint**: the amount of carbon dioxide released into the atmosphere as a result of human activity.
>
> **Global warming**: a rise in temperature of the Earth's atmosphere caused by gases and pollution.

> **Ecological deficit**: a measure which shows that more natural resources are being used than nature can replace.

How energy is generated and stored

Energy is needed:

● to manufacture products and power products and systems
● to cause something to move, heat something and create light and sound
● to process materials: extract, mould, bend, cut, drill, print and join materials.

Energy sources

Table 1.5 Non-renewable energy sources

Source	Explanation
Coal	Mined from the ground and burnt in power stations to generate electricity
Oil	Crude oil is extracted from the earth and refined into liquid fuels such as petrol. It can also generate electricity in power stations
Gas	Extracted through drilling and piped through the national grid to houses and factories. It can also generate electricity in power stations
Nuclear	Uranium ore is mined from the earth and transformed into nuclear fuel. This is used in a nuclear generator to generate heat and then converted to electricity

Table 1.6 Renewable energy sources

Source	Explanation
Wind	A turbine extracts energy from the wind. The blades are connected to a generator which produces electricity
Solar	Photovoltaic (PV) panels produce electricity when exposed to sunlight
Geothermal	Cold water is pumped underground and heated by the Earth's heat. It can be used to heat homes, or in power stations and converted to electricity
Hydroelectric	Dams house large turbines which are built to trap water. When the water is released the pressure turns the turbines which generates electricity
Wood/ biomass	Wood is chipped and used as fuel instead of burning coal. This can provide heat for homes or be used to generate electricity
	In some biomass schemes, plants such as soy are grown to produce materials which can be processed into biofuels
Wave	It is possible to harness energy from waves, although it is not widely used. In the future, tidal power offers the possibility of extracting energy from the rise and fall of the tide

Issues surrounding the use of fossil fuels

● Waste such as carbon dioxide (CO_2) and pollutants such as sulphur dioxide are emitted into the atmosphere when fossil fuels are burnt. This can cause breathing problems and contributes to global warming.
● Fossil fuels cannot be replaced and will eventually run out.
● Fossil fuels have high energy density – they hold a lot of chemical energy per kilogram of fuel, making them ideal for transportation. Electric cars currently cannot compete with the ease of using petrol (see Motor vehicles below).

Advantages and disadvantages of renewable energy sources

- Renewable sources of energy are non-polluting and considered better for the environment.
- Although biomass fuels release CO_2 as they are burned, trees are replanted, which absorb CO_2 as they grow. The process is classed as **carbon neutral**.
- Initial outlay for the equipment needed is expensive. However, following installation, it produces free energy.
- Wind and solar power depend on weather conditions. Some people find wind farms and solar panels unsightly.
- In order to build dams used to generate hydroelectric power, valleys in rural areas are flooded. This can damage the natural habitat.
- Geothermal energy units are expensive and reliant on the underground rocks being hot near the surface.
- Manufacturers are investing in renewable energy and are installing equipment to recover waste energy from various processes to heat their offices. This will reduce energy bills and demonstrates a more ethical approach to manufacturing.

> **Carbon neutral**: no net release of carbon dioxide into the atmosphere – carbon is offset.

Renewable energy sources for products

- Small solar PV panels can produce a small current to recharge a battery. Flexible solar PV panels can be found on clothing and bags which can charge a mobile phone.
- Low-powered products can be charged from a small wind generator.
- Electronic road signs are often powered by a solar PV panel.
- Clockwork wind-up mechanisms can provide a temporary source of power for mechanical or electronic products.

> **Typical mistake**
>
> Analyse or evaluate questions require depth of knowledge and understanding with sound reasoning or judgements made within the answer. You will lose marks if answers are not sufficiently detailed, are merely descriptive or if no reasoning or judgements are offered.

Energy generation and storage in a range of contexts

Motor vehicles

- Electric cars use batteries which are recharged by being plugged into an electricity source.
- Electric cars do not produce emissions, although the power to charge them comes from fossil fuels.
- Batteries take hours to fully recharge and the car's range is limited.
- Electric cars are efficient; some kinetic energy is recovered when the driver uses the brakes. This energy can then be stored in the battery.
- Electric cars are increasing in popularity as they are cheap to run and more environmentally friendly.
- Rechargeable hybrid cars offer a greater driving range but also lower emissions.

Figure 1.5 Electronic road sign powered by solar energy

Mains-powered products

- Many household appliances are charged using mains electricity.
- Products left on stand-by are a concern as they use electricity even when they are not being used.

Battery-powered products

- Energy is stored in rechargeable batteries in products such as mobile phones, tablets and cordless products.
- Solar panels absorb energy from the sun, passing the power to a battery as an electrical charge. This can power products such as garden lights.
- Some products such as torches or television remote controls are powered by non-renewable batteries, which need to be replaced.
- As batteries contain chemicals, they should be disposed of through recycling schemes.

Now test yourself

TESTED

1 Explain how automation used in industry is changing the way products are manufactured. [4]
2 Describe a situation where a consumer would be entitled to compensation from a service provider. [2]
3 Biomass fuels are said to be carbon neutral. Explain the meaning of this term. [2]
4 Describe in detail one advantage and one disadvantage of wind power. 2 x [2]
5 Explain the benefits to the environment of the 'circular economy'. [3]

2 Smart materials, composites and technical textiles

Smart materials

REVISED

Smart materials react to a change in their environment such as temperature, light pressure or electrical input. Reactions include a change in colour, shape or resistance.

Electroluminescent material

- **Electroluminescent** wire (EL wire) is made from a thin copper wire core coated in phosphor powder. It produces a glowing light when exposed to an alternating electric current.
- Electroluminescent technology is also found in flexible films or thin panels. The light-emitting phosphor is sandwiched between a pair of conductive electrodes and subjected to an AC current to create light. The brightness depends on the voltage applied.
- EL films are replacing LCD displays because they are flexible, do not generate heat and are more reliable.

Electroluminescent: materials that provide light when exposed to a current.

Quantum tunnelling composite (QTC)

- **Quantum tunnelling composites** are flexible polymers that contain conductive nickel particles that can be either a conductor of electricity or an insulator.

Quantum tunnelling composites: materials that can change from conductors to insulators when under pressure.

- The nickel particles make contact with each other and are compressed when force is applied, leading to an increase in conductivity. When the force is removed, the material returns to its original state and becomes an electrical insulator.

Shape memory alloys (SMA)

- SMAs return to their original shape if heated.
- Nitinol is a common shape memory alloy made from titanium and nickel.
- Possible uses include medical applications such as medical fastenings used in bone fractures.

Polymorph

- Polymorph is a thermoforming polymer supplied in granular form. When heated in water to 62°C it softens and forms a pliable volume of material that can be moulded and shaped.
- It solidifies on cooling and can be modelled and shaped.
- If reheated in water it becomes pliable again.
- Polymorph is a useful material for model making and prototyping.

Photochromic pigment

- Photochromic pigments or dyes change colour in response to changes in light, for example, sunglasses can change colour in response to UV radiation.

Thermochromic pigment

- Thermochromic pigments or dyes change colour in response to a change in heat and can be engineered to specific heat ranges.
- Thermochromic dyes can be used in baby bottles to give an indication of the temperature of the milk.

Micro-encapsulation

- **Micro-encapsulation** is a process of applying microscopic capsules to fibres, fabrics, paper and card.
- The capsules can contain vitamins, therapeutic oils, moisturisers, antiseptics and anti-bacterial chemicals which are released through friction.

Biomimetics

- **Biomimicry** is when inspiration for new materials, structures and systems come from the natural world.
- Fastskin, developed by Speedo, mimics the shark's sandpaper-like skin by reducing drag in the water. It is used for performance-enhancing swimwear.

1 Granules of polymorph   2 Add hot water   3 Lift out of water when soft

4 Mould to shape

Figure 2.1 The four stages of polymorph

Figure 2.2 Thermochromic mugs change colour when boiling water is poured into the mug

Micro-encapsulation: tiny microscopic droplets containing various substances applied to fibres, yarns and materials, including paper and card.

Biomimicry: taking ideas from and mimicking nature.

Composites

REVISED

Composites are created when two or more materials are joined together to create a new enhanced material. One material is known as the matrix while the other is the reinforcement.

Carbon-fibre reinforced polymer (CFRP)

- CFRP consists of woven carbon fibre strands encased in a polymer resin.
- Carbon fibre strands have a very high tensile strength and the polymer resin is lightweight and rigid. It is often used in sports equipment such as a tennis racquet where strength to weight ratio is important.

Kevlar

- Kevlar® is a lightweight, flexible and extremely durable aramid fibre that has excellent resistance to heat, corrosion and damage from chemicals and a high tensile strength-to-weight ratio.
- It is often used in protective clothing such as police body armour, where the fibre is woven in a lattice that provides protection against knife attack.

Glass reinforced plastic (GRP)

- GRP is a composite of glass fibres and a polyester resin. Glass fibres create rigidity and the resin makes GRP tough and lightweight.
- GRP is difficult to recycle because the process of combining the glass fibres and resin cannot be reversed.
- GRP is used in surfboards, canoes and car bodywork.

Technical textiles ━━━━━━━━ REVISED ▢

Technical textiles are engineered with specific performance characteristics that suit a particular purpose or function.

Interactive textiles

- **Interactive textiles** or integrated textiles are when electronic devices and circuits are integrated or embedded into textile fabric and clothing to interact and communicate with the wearer.
- Conductive fibres and threads developed from carbon, steel and silver can be woven into textile fabrics and made into clothing or conductive threads can be sewn into a product to connect a circuit.

Microfibres

- **Microfibres** are extremely fine lightweight synthetic fibres, usually polyester or nylon, that are significantly finer than a human hair.
- Useful properties include excellent strength-to-weight ratio, water-resistance and breathability.
- Fabrics made from microfibres are used throughout the textiles industry from clothing to cleaning cloths.

Phase-changing materials

- **Phase-changing materials (PCMs)** change from one state to another with the ability to absorb, store and release heat over a small temperature range by changing from liquid to solid and vice versa.

Interactive textiles: fabrics that contain devices or circuits that respond and react with the user.

Microfibres: tiny fibres about 100 times thinner than a human hair.

Phase-changing materials: encapsulated droplets on fibres and materials that change between liquid and solid within a temperature range.

Figure 2.3 Flexible solar panels can now be integrated into textile fabrics

- PCMs absorb energy during the heating process (returning to liquid) and release energy to the environment during cooling (returning to solid).
- In cold weather clothing, PCMs encapsulated into the fabric allow body heat to be stored and then released when needed.

Breathable fabrics

- Gore-Tex® consists of three or more fabrics laminated together with the breathable **hydrophilic membrane** in the middle.
- Warm air and tiny droplets of moisture from perspiration can permeate out though the breathable membrane but moisture from larger rain droplets and wind cannot enter.
- When used in high performance clothing and footwear, Gore-Tex helps to regulate body temperature by maintaining a constant temperature by allowing the flow of air in and out.

Sun-protective clothing

- Clothes made from tightly woven or knitted fabrics are the most effective at blocking out the sun's harmful UV rays because the gaps between the yarns are significantly reduced, which prevents the rays from getting through.
- Elastane fibres used in fabric reduce the spaces even further, making these types of fabric even more protective.

Nomex

- Nomex® is an **aramid** synthetic fibre primarily used where resistance to heat and flames is essential, typically in firefighters' uniforms and racing car drivers' clothing
- It is extremely strong and can withstand the most extreme conditions.

Geotextiles

- **Geotextiles** are woven or bonded, synthetic or natural, permeable fabrics made originally for use with soil with the ability to filter, separate, protect and drain.
- Geotextiles are used in civil engineering, road and building construction and maintenance, for example, control of coastal erosion and drainage and control of embankments on the sides of roads.

Figure 2.4 A geotextile is used on the roof of the Eden Project in Cornwall

Hydrophilic membrane: a solid structure that stops water passing through but at the same time can absorb and diffuse fine water vapour molecules.

Typical mistake

Technical fabrics such as Gore-Tex which appear 'breathable' should not be confused with other 'absorbent' fabrics such as cotton that simply soak up moisture and are technically not breathable.

Aramid fibre: a non-flammable heat resistant fibre at least 60 times stronger than nylon.

Geotextiles: textiles associated with soil, construction and drainage.

Exam tip

Be able to fully explain how smart, composite and technical textiles are used in the design and manufacture of products. Know their specific properties and what makes each suitable for a particular purpose in a product.

Rhovyl

- Rhovyl® is a non-flammable, synthetic fibre which is crease resistant, has good thermal and acoustic properties, is anti-bacterial and comfortable to wear.
- The construction of the fibre gives the fabric the ability to wick away moisture such as perspiration though the fabric. It also dries quickly and does not retain odours, making it ideal for socks.

Now test yourself

TESTED

1 Describe two advantages of using electroluminescent material in a product such as a lamp. [2]
2 Explain why carbon fibre reinforced polymer (CFRP) is a suitable material for a racing bike. [3]
3 Explain how micro-encapsulation is beneficial to a patient, when used in medical dressings. [2]
4 Discuss the use of polymorph as a modelling medium for product designers. [4]
5 Describe two reasons why the aramid fibre Nomex is used in protective clothing. 2 x [2]

3 Electronic systems and programmable components

- Electronic **systems** are used to provide functionality to products and processes.
- Electronic systems can be broken down into subsystems, which can be classified as inputs, processes or outputs.
- A system (or block) diagram shows how the **subsystems** are connected and the signals flow between them.
- Signals can be digital or analogue.

> **System**: a set of parts which work together to provide functionality to a product.
>
> **Subsystem**: the interconnected parts of a system.

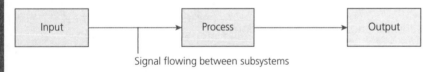

Signal flowing between subsystems

Figure 3.1 A generic electronic system

Electronic control systems

REVISED

Process subsystems

- The process subsystem receives signals from the inputs and responds in a specific way to control the outputs, depending on the needs of the product.
- Process subsystems can be made from semiconductor devices such as a **microcontroller**, microprocessor or computer.
- A microcontroller is a miniaturised computer which is programmed to perform a specific task.
- A microcontroller is an example of an **integrated circuit (IC)**. An IC is a miniaturised, highly complex circuit in a single component.

> **Microcontroller**: a miniaturised computer, programmed to perform a specific task, and embedded in a product.
>
> **Integrated circuit (IC)**: a miniaturised, highly complex circuit with small pin connections in a single component.

Table 3.1 Process devices

Process devices can perform functions such as:	Example application
Counting	Sports scoreboard, digital clock, pedometer
Switching	Night light, electric kettle, automatic door
Timing	Security light, burglar alarm, cooking timer

Inputs

- Inputs consist of sensors, which can monitor and measure a range of physical quantities.
- A sensor produces an electrical signal which can be digital or analogue.
- A **digital sensor** detects a yes/no situation, such as 'is the button pressed?'
- A switch is a digital sensor. Many types of switches are available.
- An **analogue sensor** can measure how big a quantity is, such as 'how bright is the light?' or 'what is the temperature?'

Light-dependent resistor (LDR)

- An **LDR** is an analogue sensor used to sense light.
- Its resistance falls as light level increases.
- LDRs are used in streetlamps, night lights, digital clocks (to control display brightness) and CCTV cameras (to switch to night vision mode).

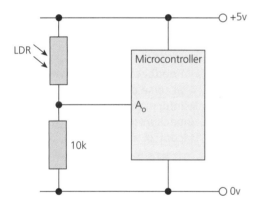

Figure 3.2 An LDR connected to an analogue input of a microcontroller

Thermistor

- A **thermistor** is an analogue sensor used to sense temperature.
- There are two types of thermistor:
  - ○ negative temperature coefficient (NTC) type: resistance **falls** as temperature increases
  - ○ positive temperature coefficient (PTC) type: resistance **rises** as temperature increases.
- Examples where thermistors are used include ovens, room thermostats, electric heaters and car engines.

> **Digital sensor**: a sensor to detect a yes/no or an on/off situation.
>
> **Analogue sensor**: a sensor to measure how big a physical quantity is.
>
> **Light-dependent resistor (LDR)**: an analogue component to sense light level.

> **Typical mistake**
>
> Confusing digital and analogue sensors and signals is a common error. Make sure you know which sensor belongs to which category.

> **Thermistor**: an analogue component to sense temperature.

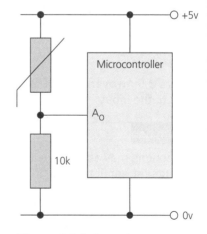

Figure 3.3 A thermistor providing an input signal to a microcontroller

Outputs

- An output subsystem converts an electrical signal into a desired function.
- A buzzer produces a sound output. Buzzers are useful for providing feedback that a user has pressed a button. They are found in many products including burglar alarms, microwave ovens, dishwashers and kitchen timers.
- A light-emitting diode (LED) produces a light output. They are available in a range of colours, sizes and shapes and can be used as an indicator, for example, as a 'power on' light in a product, or as a source of illumination, for example, in a torch.
- A resistor must be used with an LED to limit the current flowing, or the LED will burn out.

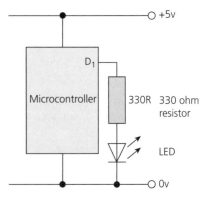

Figure 3.4 **Using an LED as an output from a microcontroller**

> **Typical mistake**
>
> Incorrectly describing how the resistance of LDRs and thermistors (NTC and PTC types) changes with light and temperature.

Feedback in control systems

- **Feedback** in a system is when a signal from the output is taken and fed back into the input of the process subsystem.
- Feedback allows a microcontroller to monitor the effect of the changes it makes to its output devices in order to achieve precise control over a system.
- An example of feedback is in an electric oven control system.

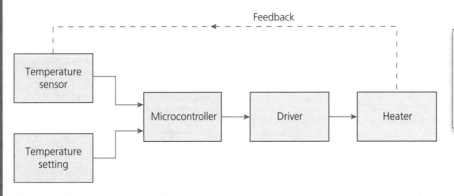

> **Feedback**: achieving precise control by feeding information from an output back into the input of a control system.

Figure 3.5 **An oven control system, where a signal from the output is fed back to the input**

> **Exam tip**
>
> Questions may ask you to identify suitable input/output devices for a particular application, or to classify devices into digital and analogue types. You may be asked to complete a system diagram for a given application, so practise drawing system diagrams for several familiar products, including systems with feedback.

> **Typical mistake**
>
> Forgetting to use a resistor with an LED.

Programmable systems

Microcontrollers

- A programmed microcontroller is embedded into a product to provide functionality and to enhance performance.
- Microcontrollers can be reprogrammed, which is useful during product development or for product upgrades.
- Microcontrollers are found in products including toasters, TVs, microwave ovens, hi-fi systems and cars.
- Microcontrollers can be interfaced with a wide range of digital and analogue input and output devices.
- Some output devices require a driver to boost the output signal from the microcontroller.
- A **programmable interface controller (PIC)** is a popular type of microcontroller. PICs are used in many GCSE projects.

Flowchart programs

- A microcontroller **program** is a set of instructions which tells the microcontroller what to do.
- When a program is run, the microcontroller executes the instructions at extremely high speed.
- A **flowchart** is a graphical way of showing a program.
- Standard symbols are used in flowchart programs.

Examples of flowchart commands include:

- **input/output** commands: 'Read the temperature sensor', 'Turn on LED', 'Turn off buzzer'
- **process** commands: 'Wait 2 seconds', 'Add 1 to the value of variable A'
- **decision** commands: 'Has temperature dropped below 5°C?', 'Is the button pressed?', 'Is A>45?'.

Exam tip

Questions on programmable devices may require you to draw (or to complete) a flowchart program, so learn the flowchart symbols and practise drawing flowcharts for the control of familiar products.

Subroutines

- **Subroutines** (also called 'macros') can be used to help simplify the structure of a complex program.
- A subroutine is a set of program instructions that performs a specific task, for example, 'flash an LED five times', or 'measure how long a user holds a button pressed'.
- Subroutines are called from the main program using a Call command.
- A Return command at the end of the subroutine returns the flow back to the main program.

Programmable interface controller (PIC): A microcontroller used in many products.

Program: a set of instructions which tells the microcontroller what to do.

Flowchart: a graphical representation of a program.

Symbol	Name
	Start/end
→	Arrows
	Input/output
	Process
	Decision

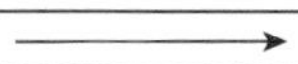

Figure 3.6 Flowchart symbols

Subroutine: a small sub-program within a larger program.

Typical mistake

Using the wrong flowchart symbols for input/output, process and decision commands.

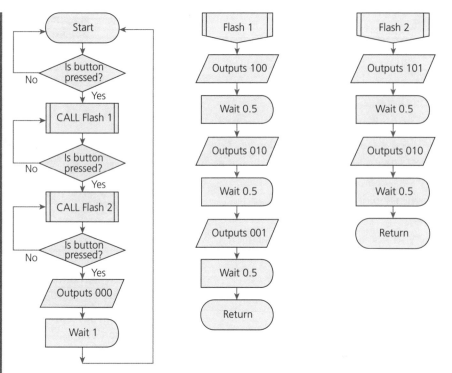

Figure 3.7 A flowchart program using two subroutines

4 Mechanical components and devices

Types of motion

REVISED

There are four types of **motion**:

- **rotary motion** – for example, wheels and electric motors
- **linear motion** – for example, cars and conveyor belts
- **oscillating motion** – for example, an electric toothbrush head or a pendulum
- **reciprocating motion** – example, a needle on a sewing machine or a jigsaw blade.

Rotary motion:

$$\text{Rotational speed} = \frac{\text{number of revolutions}}{\text{time taken}}$$

Linear motion:

$$\text{Speed} = \frac{\text{distance travelled}}{\text{time taken}}$$

> **Motion:** when an object moves its position over time.
>
> **Rotary motion:** movement in a circular path.
>
> **Linear motion:** movement in a straight line.
>
> **Oscillating motion:** movement back and forth in a circular path.
>
> **Reciprocating motion:** movement back and forth in a straight line.

Oscillating and reciprocating motion:

$$\text{Frequency (oscillation speed)} = \frac{\text{number of oscillations}}{\text{time taken}}$$

- Rotational speed is often measured in 'revolutions per minute (rpm)' or, sometimes, 'revolutions per second'. To convert revolutions per second into rpm, multiply by 60.
- Speed is measured in various units, and you should to look carefully at the units given in an exam question. Typical units are metres per second (ms^{-1}), kilometres per hour ($km\ h^{-1}$) or millimetres per second ($mm\ s^{-1}$).
- To convert $mm\ s^{-1}$ into ms^{-1}, divide by 1000.

Mechanical systems REVISED

- Mechanical systems can produce different types of movement.
- Mechanical systems can change the magnitude (size) and direction of forces and movement.
- A mechanical system will take an input force (or motion) and process it to produce an output force (or motion).
- A **force** is a push, a pull or a twist.
- A simple **mechanism** trades-off forces against distances moved. If one increases, the other must decrease.

Mechanical components REVISED

Levers

- A **lever** is a rigid bar that pivots on a **fulcrum**. The input force is called the **effort**, and the output force is called the **load**.
- The lever arm length is the distance between the force and the fulcrum.
- If the input arm length is larger than the output arm length, the lever will increase the force applied, but reduce the distance moved by the force. In other words, the load will be larger than the effort.
- The lever arm length is in inverse proportion to the force. If the output arm is half as long as the input arm, the output force will be twice as big as the input force.

$$\frac{\text{Load}}{\text{Effort}} = \frac{\text{input arm length}}{\text{output arm length}}$$

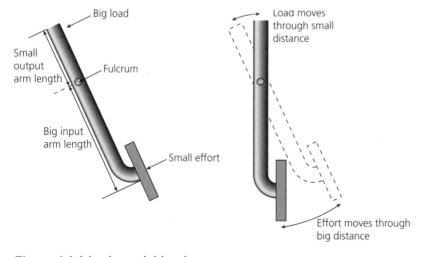

Figure 4.1 A brake pedal is a lever

> **Exam tip**
>
> Make sure you can use, and rearrange, the speed – distance – time formula.

> **Force**: a push, a pull or a twist.
>
> **Mechanism**: a series of parts that work together to control forces and motion.

> **Exam tip**
>
> Questions may ask you to sketch examples of mechanisms, using arrows to indicate movement, so practise sketching a variety of mechanisms and mechanical components.

> **Lever**: a rigid bar that pivots on a **fulcrum**.
>
> **Fulcrum**: the pivot point on a lever.
>
> **Effort**: the input force on a lever.
>
> **Load**: the output force from a lever.

Linkages

- A **linkage** is used to direct forces and movement to where they are needed.
- A **simple pulley** changes the direction of motion of a cord.
- A **reverse motion linkage** will reverse the direction of input motion.
- A **bell crank** will transfer motion around a corner.
- A **peg and slot** converts rotary motion into oscillating motion.
- A **crank and slider** converts rotary motion into reciprocating motion.

> **Linkage**: a component to direct forces and movement to where they are needed.

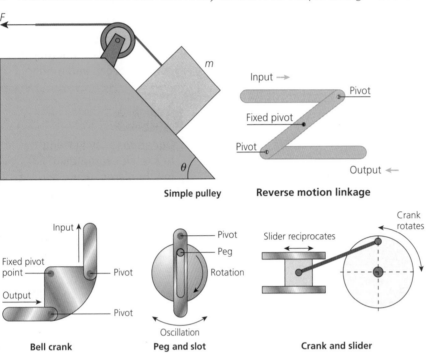

Figure 4.2 Examples of linkages

Cams

- A **cam** and follower converts rotary motion into reciprocating motion.
- A **snail cam** causes the follower to steadily rise, followed by a sudden drop.
- A **pear-shaped cam** creates a sudden rise and fall followed by a long period where the follower does not move.
- The **eccentric cam** creates an even rise and fall motion throughout its rotation.
- Cams are used in toys, machinery and engines.

> **Cam**: a component used with a follower to convert rotary motion to reciprocating motion.

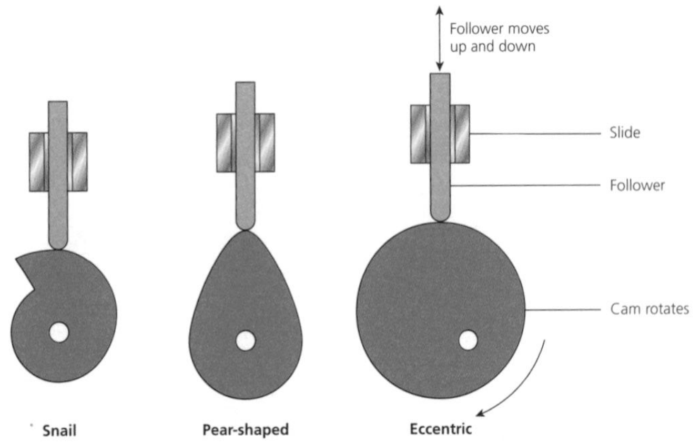

Figure 4.3 Three types of cam

Gears

- Gear systems transfer rotary motion.
- A **simple gear train** consists of two **spur gears**, which are wheels which interlock with teeth around their edge.
- The input gear is called the **driver gear**. The output gear is called the **driven gear**.
- If the driver is the smaller gear, it is sometimes called a pinion.
- The two gears will rotate in opposite directions.
- The smaller gear will rotate faster than the larger gear.
- The number of teeth on the gear is in inverse proportion to the speed it rotates. A gear with twice as many teeth will rotate at half the speed.

$$\frac{\text{Input gear speed}}{\text{Output gear speed}} = \frac{\text{number of teeth on driven gear}}{\text{number of teeth on driver gear}}$$

- Gear systems are used in cordless drills, clocks, winches and cars.

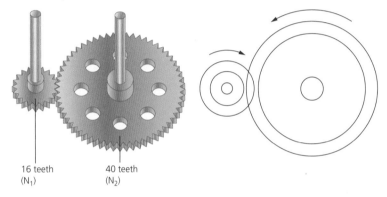

16 teeth (N_1)   40 teeth (N_2)

Figure 4.4 A simple gear train

Belt drives

- A **pulley and belt** drive transfers rotary motion, like a gear system.
- The input and output pulleys can be separated by a large distance by using a long belt.
- The smaller pulley will rotate faster than the larger pulley.
- The input and output pulleys rotate in the same direction.
- Belt drives are used in washing machines, workshop bench drills and car engines.

Rack and pinion

- A rack and pinion converts rotary motion into linear motion.
- Examples of uses of rack and pinion mechanisms are stairlifts and sliding doors.

Figure 4.6 A rack and pinion

Simple gear train: two spur gears meshed together.

Spur gear: a gear wheel with teeth around its edge.

Driver gear: the input gear on a gear train.

Driven gear: the output gear from a gear train.

Typical mistake

Confusing the directions of motion in lever systems and simple gear trains, and confusing whether these systems increase or decrease the magnitude of forces and of movement.

Figure 4.5 A belt drive in a washing machine

Exam tip

Questions may test your knowledge and understanding of how linkages and mechanisms change the type of motion, and how they change the magnitude and direction of forces, so make sure you understand this for the various mechanisms.

Now test yourself

1 Identify the four types of motion, giving an example for each type. [4]
2 A wheel rotates at 300 rpm. Calculate the number of revolutions it will make in 10 s. [2]
3 A robot travels forwards at a speed of 0.8 ms^{-1}. Calculate the time taken for it to travel 5 m. [3]
4 Draw a diagram of a lever which will double the size of an applied effort force. Label the effort, load and fulcrum. [3]
5 Draw labelled sketches of the following mechanisms and identify the conversion of motion that takes place in each one:
   (a) crank and slider   (b) rack and pinion   (c) bell crank. [6]
6 Give two functional similarities and two functional differences between a simple gear train and a belt and pulley drive system. [4]

5 Materials and their working properties

Papers and boards

Categorising paper and boards

- Paper and board is available in standard sized sheets (see Figure 5.1).
- Each sheet size is twice the size of the one before, for example, A3 is twice the size of A4.
- If we fold a sheet of paper in half it then becomes the next size below. For example, an A1 sheet folded in half becomes A2 size.
- The thickness of paper is known as its 'weight'.
- Weight is measured in **grams per square metre** (g/m^2 or gsm).
- A weight greater than 170 gsm is classified as a board rather than a paper.
- Boards are usually classified by thickness as well as by weight.
- The thickness of board is measured in **microns** (one-thousandth of a millimetre).

> **Grams per square metre (gsm):** the weight of paper and card.
>
> **Micron:** one thousandth of a millimetre (0.001 mm) – used to measure the thickness of board.

Paper

Size	A10	A9	A8	A7	A6	A5	A4	A3	A2	A1	A0	2A0	4A0
Length (mm)	37	52	74	105	148	210	297	420	594	841	1189	1682	2378
Width (mm)	26	37	52	74	105	148	210	297	420	594	841	1189	1682

most common sizes used by designers

Figure 5.1 Paper sizes

Table 5.1 Common paper types

Paper type	Properties	Uses
Layout paper	Smooth surface Weighs around 50 gsm Thin enough to trace and copy parts of designs Cheap	Sketching and developing design ideas
Tracing paper	Thin and very transparent Weighs around 40 gsm Hardwearing and strong Mistakes in pen can be scratched off using a sharp blade	Making copies of drawings and fine details
Copier paper	Smooth surface Weighs approximately 80 gsm	Printing, photocopying and general office purposes
Cartridge paper	Available in different weights between 80–140 gsm More expensive than layout and copier paper Has a slightly textured surface and is slightly creamier in colour Ideal surface for pencil, crayons, pastels, watercolour paints, inks and gouache	Sketching, drawing and painting

Virgin fibre paper

- Paper and card are made from wood fibres called cellulose.
- Paper made from 'new' wood fibres is called **virgin fibre paper**.
- Chemicals are used to break down or 'pulp' the fibres.
- Further chemicals can be added to bleach or colour the paper and give it a special texture.

Recycled paper

- **Recycled paper** is made by soaking waste paper to separate the fibres back into pulp.
- The pulp is refined to remove unwanted contaminants such as staples and plastic.
- Each time paper is recycled the fibres get shorter and weaker.
- Most recycled paper is a mixture of virgin pulp and pulp from recycled paper.
- The higher the proportion of recycled pulp, the lower the quality of paper produced.

> **Virgin fibre paper**: paper made entirely from 'new' wood pulp.
>
> **Recycled paper**: paper made from wood pulp using some re-pulped paper.

Table 5.2 Composition and uses of recycled paper

Virgin pulp	Pulp from recycled fibres	Products and uses
85% to 100% (approx.)	0% to 15% (approx.)	High-quality printing and graphic paper, publication paper
70% to 85% (approx.)	15% to 30% (approx.)	Household and sanitary applications (e.g. toilet paper, kitchen roll)
50% to 70% (approx.)	30% to 50% (approx.)	Newspaper
20% to 50% (approx.)	50% to 80% (approx.)	Paper and cardboard packaging

Card and carboard

Table 5.3 Card and cardboard

Board type	Properties	Uses
Card	Around 180 to 300 gsm in weight Wide range of colours, sizes and finishes Easy to fold, cut and print on	Greetings cards, paperback book covers and simple models
Cardboard	Around 300 microns upwards in thickness Inexpensive and can be cut, folded and printed on easily	Packaging, e.g. cereal boxes, tissue boxes, sandwich packets Modelling design ideas and making templates for parts made from metal or other resistant materials
Folding boxboard	Similar in thickness to cardboard but more rigid and lightweight Made of layers of mechanical pulp sandwiched between two thinner layers of chemical pulp A coating is applied to one side to give it a smoother texture and white colour	Packaging for products such as frozen foods, medicines and beauty products
Corrugated cardboard	A strong but lightweight type of card Made from two layers of card with a fluted sheet in between Available in thicknesses ranging from 3 mm (3000 microns) upwards	Used for packaging fragile or delicate items that need protection during transportation Used as packaging for takeaway food due to good heat-insulating properties
Board sheets	Mounting board is a rigid type of card around 1.4 mm thickness (1400 microns) Available in different colours but white and black are the most common	Picture framing mounts and architectural modelling

Laminating

There are two different types of lamination.

- Coating surfaces during the paper-making process to reduce absorbency, give clearer printing and increase strength.

- Applying a thin film of clear plastic to one or both sides of paper or thin card. Examples include menus, posters, signs, identity badges. It improves strength and resistance, waterproofs the document, improves the appearance and increases the lifespan of the printed document.

Different methods of laminating a document are covered in 'Topic 7 Paper and boards'.

Natural and manufactured timber

REVISED

- Timber is recyclable, reusable and renewable.
- Timbers are readily available as planks, boards, strips, dowels, mouldings and square sections.

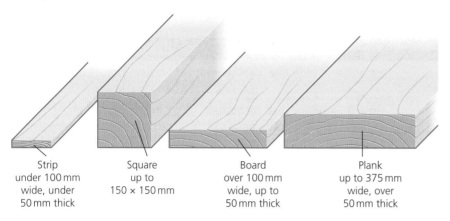

Strip
under 100 mm
wide, under
50 mm thick

Square
up to
150 × 150 mm

Board
over 100 mm
wide, up to
50 mm thick

Plank
up to 375 mm
wide, over
50 mm thick

Figure 5.2 Timber stock forms

Natural timber

Hardwoods

- **Hardwoods** come from deciduous trees. Most deciduous trees lose their leaves in autumn. Deciduous trees take a long time to mature.
- Hardwoods generally have a close grain structure. This usually makes them harder and stronger than softwoods.
- They can be sanded to a finer, smoother finish.
- Hardwoods are generally more expensive than softwoods.

Hardwood: timber that comes from deciduous trees.

Softwoods

- **Softwoods** come from coniferous trees, also known as evergreens.
- Coniferous trees are quick growing and take around ten years to reach maturity.
- Most softwoods have an open grain and are generally less dense and not as strong as hardwoods.

Softwood: timber that comes from coniferous trees.

Table 5.4 Examples of common hard and softwoods

Type	Properties	Uses
Hardwoods e.g. oak, beech, maple	Hardwearing, heavy, attractive grain and smooth finish	Furniture, children's toys and outdoor applications
Softwoods e.g. Scots pine, spruce	Inexpensive, easy to cut, shape and smooth	Floorboards and roof trusses

Manufactured boards

- **Manufactured boards** fall into two categories: **laminated boards** and **compressed boards**.
- Laminated boards are produced by gluing large sheets or **veneers** together.
- Compressed boards are manufactured by gluing particles, chips or flakes together under pressure. They are often covered with thin polymer laminate.
- Manufactured boards are available in large sheets. They are less expensive than natural timbers.

Manufactured boards: man-made boards that are available in large sheets.

Laminated boards: manufactured boards made by layering sheets together.

Compressed boards: manufactured boards made by gluing wood particles together.

Veneer: thin sheet of natural timber used to cover manufactured boards.

Table 5.5 Examples of manufactured boards

	Properties	Uses
Laminated boards e.g. plywood, blockboard, veneered board	Strong and look like 'real' wood	Shelving, workbenches and worktops
Compressed boards e.g. medium-density fibreboard (MDF), chipboard, hardboard	Smooth surface, can be coated with plastic laminate, easy to cut and shape and no grain	Kitchen worktops, cupboards and bedroom furniture

Finishes

- Finishes help protect timber from damage and enhance its appearance.
- Wood stains change the colour of the timber but don't add protection.
- Paint helps protect wood from the weather.
- Varnish is a clear coating that gives protection against weather and enhances the appearance.
- Oils such as Danish oil and teak oil give timber an improved appearance and add a low level of protection.
- Wood preservatives used on outdoor products protect from weather and help prevent decay.
- Manufactured boards can be varnished, stained and painted but a coating must first be applied to seal the porous surface of the board.

Ferrous and non-ferrous metals

REVISED

- Metal is a naturally occurring material mined from the ground in the form of ore.
- The raw metal is extracted from the ore by crushing, smelting or heating.
- Metals are available in a variety of stock forms such as sheet, rod, bar, tube and angle.

Ferrous metals

- **Ferrous metals** are those that contain iron.
- Steel is the most widely used form of ferrous metal. It is easy to recycle.
- Most ferrous metals are magnetic due to the iron content.
- They are prone to corrosion when exposed to moisture and oxygen.
- The properties of ferrous metals, such as hardness and malleability, are directly related to their carbon content – the more carbon, the harder and less malleable the steel becomes.

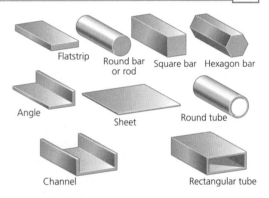

Flatstrip Round bar or rod Square bar Hexagon bar

Angle Sheet Round tube

Channel Rectangular tube

Figure 5.3 Metal stock forms

> **Ferrous metals**: metals that contain iron.

Table 5.6 Examples of ferrous metals

	Properties	Uses
Mild steel	Good tensile strength Malleable (easy to cut and shape)	Rolled steel joists (RSJs), car body panels and office furniture
Medium-carbon steel	High strength, but less malleable	Gardening tools such as spades and trowels
High-carbon steel	Very high strength but even less malleable	Cutting tools such as saw blades and drill bits
Cast iron	Hard but brittle	Metalwork vices and cookware

Non-ferrous metals

- **Non-ferrous metals** do not contain iron and do not corrode.
- Most are not magnetic which makes them ideal for use in electronic devices and wiring.
- After steel, aluminium is the most widely used and recycled metal.
- Aluminium takes a huge amount of energy to extract and manufacture. It takes around 95 per cent less energy to recycle aluminium than to produce the raw material.

Non-ferrous metals: metals that do not contain iron.

Table 5.7 Examples of non-ferrous metals

	Properties	Uses
Aluminium	Extremely lightweight, soft and malleable	Drinks cans, foils, kitchen utensils and aeroplane parts
Copper	Good electrical conductivity, good thermal conductivity and corrosion resistance	Plumbing pipes, electrical wire and roofing
Lead	Heavy, soft, highly malleable and ductile	Fishing and diving weights, roofing and batteries

Alloys

- **Alloys** are metals combined with other metals or elements to improve the properties.
- Alloys can be ferrous or non-ferrous depending on the main pure metal that they contain.

Alloy: a mixture of two or more metals.

Typical mistake

Don't confuse recyclability with sustainability! Even though metal can be recycled over and over again, it is not a sustainable material. Although timber, paper and board are sustainable materials, they can only be recycled a few times

Table 5.8 Examples of alloys

	Elements	Properties	Uses
Stainless steel (ferrous alloy)	Iron Carbon Manganese Chromium	Extremely hard Corrosion resistant Bright shiny finish	Surgical equipment Cutlery
Brass (non-ferrous alloy)	Copper Zinc Lead	Easy to cast Lustrous appearance Low coefficient of friction	Locks, gears, bearings Ammunition casings Musical instruments
Bronze (non-ferrous alloy)	Copper Tin	Good resistance to corrosion and fatigue Good electrical conductivity High elasticity	Coins Ship propellers Submerged bearings

Finishes

- Ferrous metals require a finish to be applied to prevent them rusting.
- Ferrous metals can be painted, plated, galvanised, powder coated or polymer coated.

- Non-ferrous metals do not rust but will discolour due to oxidisation.
- Chrome plating, anodising and coating with clear lacquer can be used to finish non-ferrous metals.

Thermoforming and thermosetting polymers

- **Polymers** can be coloured, shaped and formed to produce products.
- They are often used to improve the performance of products and replace more traditional materials.
- Polymers can be identified by their weight, hardness, elasticity, conductivity/insulation, toughness, strength and mouldability.
- Most polymers are synthetic polymers and are made from crude oil which is an unsustainable resource. Chemical engineers are looking for reliable alternatives.
- **Natural polymers** (known as biopolymers) come from renewable sources such as corn starch.
- Natural polymers are relatively new and there are far fewer types.
- Polylactate acid (PLA) is the most common biopolymer; it is used in 3D printers.
- Polymers are readily available in a number of different stock forms including sheet, film, bar, rod, granules, powder and tube.

> **Polymer**: substance or fibre with a molecular structure made up of smaller units bonded together.
>
> **Natural polymer**: polymer made from natural sources.

Figure 5.4 Acrylic (PMMA) sheet

Thermoforming polymers

- **Thermoforming polymers** can be softened by heating and moulded into almost any shape.
- Once softened, or plasticised, they can be shaped and formed using a wide variety of processes such as bending, vacuum forming, moulding and extrusion.
- Once the desired shape has been achieved the polymer cools and maintains its new shape.
- Thermoforming polymers can be re-heated, re-shaped and cooled many times with minimal damage to the properties of the polymer.
- Thermoforming polymers can be recycled.

> **Thermosetting polymer**: polymer that can only be softened by heating and shaped once.
>
> **Thermoforming polymer**: polymer that can be softened by heating, re-shaped and set over and over again.

Thermosetting polymers

- **Thermosetting polymers** cannot be reheated or reformed once they have been formed and cooled.
- Thermosetting polymers make excellent insulators.
- Thermosetting polymers cannot be recycled.

Table 5.9 Examples of thermoforming and thermosetting polymers

	Type of polymer	Properties	Uses
Polypropylene (PP)	Thermoforming	Semi-rigid, good chemical resistance, tough, good fatigue resistance, integral hinge property, good heat resistance	Buckets, bowls, crates, toys, bottle caps, car bumpers and jug kettles
Polythene (PE)	Thermoforming	Tough, flexible, easily moulded	Carrier bags, bin liners, washing up bottles
Polystyrene (PS)	Thermoforming	Lightweight, food safe, good impact strength	Disposable plastic cutlery, CD cases, smoke detector housings and plastic model assembly kits
Polyvinyl chloride (PVC)	Thermoforming	Rigid, dense, good tensile strength	Water pipes, window frames, raincoats and flooring
Melamine formaldehyde (MF)	Thermosetting	Tasteless, odourless, shrink resistant, chemical resistant, heat resistant, hard, scratch and impact resistant	Household crockery items (glasses, cups, bowls and plates), toilet seats and pan knobs and handles
Urea formaldehyde (UF)	Thermosetting	Hardwearing, durable, non-conductor of electricity, lower water absorption, excellent heat resistance	Electrical fittings, resin used in wood adhesives
Polyester resin (PR)	Thermosetting	Rigid, brittle, good electrical insulator, good chemical resistance	Moulds for casting, boat hulls and car body shells and panels
Epoxy resin (ER)		Rigid, brittle, thermosetting, good electrical insulator, good chemical resistance	Moulds for casting, adhesives and circuit boards

Natural, synthetic, blended and mixed fibres, and woven, non-woven and knitted textiles

REVISED ☐

- **Fibres** are very fine hair-like structures that are spun (or twisted) together to make yarns.
- Yarns are then woven or knitted together to create textile fabrics.
- Fibres (also known as polymers) have different properties and characteristics that affects what they can be used for.

Natural polymers

- Natural polymers come from natural sources – plants (**cellulosic**) and animals (**protein**).
- They are sustainable and biodegradable.
- The sources of natural fibres are:
  - plant polymers – extracted from the stem or seeds of plants, and cellulose-based fibres extracted from, for example, wood pulp
  - insect polymers – extracted from insects
  - animal polymers – fibre from the hair or fleece of animals such as sheep, goats (mohair, cashmere), rabbit (angora), alpaca and camel.

> **Fibre**: a fine hair-like structure.
>
> **Cellulosic fibres**: natural fibres from plant-based sources.
>
> **Protein fibres**: natural fibres from animal-based sources.

Table 5.10 Properties and common uses of natural polymers

Fibre	Source	Properties	Uses
Cotton	Plant	Absorbent, strong, cool to wear, hard wearing, creases easily, smooth, easy to care for, flammable, can shrink	Clothing, sewing and knitting threads, soft furnishings
Linen	Plant	Strong, cool to wear, absorbent, hard wearing, creases very easily, has a natural appearance, handles well, flammable	Lightweight summer clothing, soft furnishings, table linen
Hemp	Plant	Absorbent, non-static, anti-bacterial, naturally lustrous, strong	Clothing, carpets and rugs, ropes, mattress filling
Jute	Plant	Very absorbent, high tensile strength, anti-static	Bags, sacking, carpets, geo textiles, yarn and twine, upholstery, clothing but to a lesser extent
Silk	Insect	Absorbent, comfortable to wear, can be cool or warm to wear, strong when dry, has a natural sheen, creases	Luxury clothing and lingerie, knitwear, soft furnishings
Wool	Animal	Warm, absorbent, low flammability, good elasticity, crease resistant	Warm outerwear including coats, jackets and suits, knitwear, soft furnishings including carpets and blankets
Viscose	Plant/ chemical	Blends well with other fibres, breathable, drapes well, excellent colour retention, highly absorbent, relatively light	Linings, shirts, shorts, coats, jackets and other outerwear
Rayon	Plant/ chemical	Comfortable, soft to the skin, moderate strength and abrasion resistance	Blouses, jackets, sportswear and dresses

Manufactured polymers

- Manufactured or **synthetic** polymers are artificial fibres derived from oil, coal, minerals or petrochemicals.
- The fibres (known as monomers) are joined together by **polymerisation** then spun into yarns before being woven or knitted into fabrics.
- An advantage of synthetic fibres and yarn is that they can be engineered for specific purposes.
- Most synthetic polymers are non-biodegradable and from unsustainable sources.

> **Synthetic:** derived from petrochemicals or man-made.
>
> **Polymerisation:** chemical reaction that causes many small molecules to join together and form a larger molecule.

Table 5.11 Properties and common uses of synthetic polymers

Fibre	Properties	Uses
Polyester	Strong when wet and dry, flame resistant, thermoplastic, hard wearing, poor absorbency	A very versatile fabric used throughout textiles
Nylon (polyamide)	Strong and hard wearing, melts as it burns, thermoplastic, good elasticity, poor absorbency	Clothing, carpets and rugs, seat belts and ropes, tents
Polypropylene	Thermoplastic with a low melting point, strong, crease resistant, non-absorbent, resistant to chemicals, hard wearing and durable	Engineered for specific uses to include carpet backing, sacks, webbing, twine, fishing nets, ropes, some medical and hygiene products, awnings, geotextiles

Fibre	Properties	Uses
Acrylic	Strong except when wet, thermoplastic, burns slowly then melts, poor absorbency	Knitwear and some knitted fabrics, fake fur products including toys, upholstery
Elastane, Lycra®	Very elastic and stretchy, lightweight, strong and hardwearing	Clothing, particularly swimwear and sportswear where stretch, comfort and fit is critical
Aramid fibres	Engineered for strength and heat resistance, no melting point, five times stronger than nylon, resistant to abrasion, low shrinkage, ease of care	Flame resistant clothing, protective clothing, accessories, armour, geotextiles, aeronautical industry, ropes and cables, high risk sports equipment

Microfibres

- **Microfibres** can be natural or manufactured.
- They are 60-100 times finer than a human hair.
- Microfibres can be engineered to create fabrics with specific qualities and functions. For example, lightweight, strong, crease resistant and soft.
- Products made from microfibres include sportswear, underwear and high-performance garments.

> **Microfibre**: an extremely fine, specially engineered fibre.

Table 5.12 Examples of microfibres

Microfibre	Source	Properties	Uses
Tactel® Polyamide (nylon) fibre	Manufactured	Hardwearing, quick drying, crease resistant	Often blended with cotton or linen, used mainly for underwear and active wear
Tencel® (lyocell) Regenerated cellulose wood fibre	Natural	Biodegradeable, strong, soft, absorbent, resists creasing	Shirts, suits, skirts, leggings

Blending and mixing fibres

Fibres are often mixed or blended together to improve the properties of the yarn or fabric. For example:

- to improve the quality to make it stronger or easier to care for
- to improve the appearance
- to improve functionality, for example, to improve the handle of the fabric so that it drapes better
- to improve the cost of the **yarn** or fabric, for example, by blending an inexpensive yarn with an expensive yarn.

> **Yarn**: spun thread used for knitting, weaving or sewing.

Mixed fibres

- Fibres are mixed by adding yarns of different fibres together during the production of the fabric.
- One yarn is used for the warp yarns that run along the length of the fabric and a different one for the weft yarns that are combined with the warp yarns across the fabric.

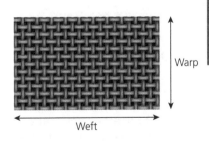

Figure 5.5 In a mixed fibre fabric, the warp yarns would be one fibre and the weft yarns would be a different fibre

Blended fibres

- Blended fibres are two or more different fibres spun together to make a single yarn.
- The most common blend is polyester cotton.
- Cotton is absorbent, soft and strong.
- Polyester is hardwearing, quick drying and elastic.
- The combined properties create a versatile fabric which is comfortable and cool to wear like cotton, but with the added features of being quick drying and crease resistant.

Woven, non-woven and knitted textiles

Woven fabrics

- Woven fabrics consist of warp and weft yarns in an under/over configuration.
- Different **weaves** create fabrics with different textures, patterns and strength.
- Plain weave is the most basic weave but is stable, strong and gives an even surface on both sides of the fabric.
- Twill weave (used for denim jeans) gives a diagonal pattern and adds strength to the cotton fabric.

Non-woven

- Non-woven textiles are constructed from a web of fibres, held together by adhesive or stitching.
- Most non-woven fabrics are not considered strong enough to be made into garments but are often used to reinforce woven and knitted fabrics.
- Non-woven fabrics are used to make disposable products, such as surgical gowns and masks, cleaning wipes and protective suits for crime scene investigators.

Knitted fabrics

- Knitted fabrics are made by creating a series of loops in the yarns that interlock together.
- Knitted fabric is easy to stretch and warmer to wear as the loops trap body heat.

> **Weave**: the pattern woven in the production of fabric.

Figure 5.6 Different coloured yarns are used to create stripes and checks

> **Exam tips**
>
> By making things in school or at home using different types of materials, you will learn their different characteristics and properties.
>
> Look at the materials used in everyday products such as packaging, furniture, electronic gadgets and vehicles to help you remember their properties.
> - Think about what material was used for the parts of the product.
> - What properties does the material have?

Now test yourself

TESTED ☐

1. Describe three reasons for laminating paper and card products. [2]
2. Explain why hardwood is considered by some to be an unsustainable material. [6]
3. Give one advantage of thermoforming polymers compared to thermosetting polymers. [2]
4. Give one example of how a metal can be tested to find out if it is ferrous or non-ferrous. [2]
5. What is a microfibre? [3]

Exam practice questions

1 (a) New products are designed and developed as a result of market pull and technology push. Explain how market pull and technology push has impacted on the design and development of fitness trackers. [4]
  (b) Describe how studying a product's life cycle benefits a manufacturer. [3]
2 Sustainability and environmental issues are important considerations when developing new products.
  (a) Define the term 'sustainable design'. [2]
  (b) Explain how the circular economy benefits the environment. [4]
3 Smart materials can change their properties or appearance in response to external stimuli.
  (a) State what makes shape memory alloys (SMAs) different to other metals. [1]
  (b) Complete the following sentence: *Nitinol is a one of the most common shape memory alloys made up of .... and ....* [2]
  (c) Describe an example where thermochromic pigment used in a named product could benefit the user. [3]
4 Electronic devices and circuits can be embedded into textile fabrics for clothing.
  (a) Heart rate monitors can be embedded into fabric. Describe how this would benefit a user with a heart condition. [2]
  (b) Micro-encapsulation is often used in medical textiles. Give two examples of its use in medicine and explain the benefit to the patient. 2 x [3]
  (c) Explain why Rhovyl® is used in sportswear. [2]
5 A coffee machine uses a negative temperature coefficient (NTC) thermistor to sense the temperature of the hot water and a microcontroller to control the function of the machine.
  (a) Complete this sentence: *As the temperature of a NTC thermistor rises, its resistance ....* [1]
  (b) Sensors can be classified into analogue or digital types. Explain why a thermistor is an analogue sensor. [2]
  (c) Describe **two** advantages to a designer of using a microcontroller in a product. [4]
  (d) The temperature control system for the coffee machine uses feedback.
    (i) Explain what is meant by feedback in a control system. [2]
    (ii) Give one advantage of using feedback in a control system. [1]
6 A toy car is propelled by an electric motor. The motor produces rotary motion at high speed.
  (a) Apart from rotary motion, identify **two** other types of motion. [2]
  (b) Draw a labelled diagram to show a mechanical system which would reduce the rotary speed produced by the motor. [3]
  (c) Describe **two** functional reasons for using levers in mechanical systems. [2]
7 Describe what happens to the fibres in paper each time it is recycled and how this affects the recycled paper's properties and uses. [3]
8 Explain the process of laminating and why this is done to certain paper and card products. [3]

ONLINE

2 In-depth knowledge and understanding

6 Electronic systems, programmable components and mechanical devices

Sources, origins, physical and working properties

REVISED

Functions of electronic and programmable systems

Basic electrical concepts

- The **voltage** at a point in a circuit is a measure of the electrical 'pressure'.
- The **current** is a measure of the actual electricity flowing.
- Voltage is measured in volts (symbol V).
- Current is measured in **amps** (symbol A) or **milliamps** (mA). 1A = 1 000mA
- **Ohm's law** explains the relationship between voltage and current. It can be summarised by the formula:

  voltage = current × resistance

  $V = IR$

- **Resistance** is measured in **ohms** (symbol Ω) or **kilohms** (kΩ) or **megaohms** (MΩ)

  1 kΩ = 1 000 Ω

  1 MΩ = 1 000 kΩ = 1 000 000 Ω

> **Voltage:** the electrical 'pressure' at a point in a circuit, in volts (V).
>
> **Current:** a measure of the actual electricity flowing, in amps (A).

Resistors

- Resistors are only made in certain resistance values, called **preferred values**.
- The E12 series is a set of 12 resistor values. The values will be provided for you in the examination, but for reference the values are:

  1.0   1.2   1.5   1.8   2.2   2.7   3.3   3.9   4.7   5.6   6.8   8.2

- Resistors are available in these values and their multiples of 10. For example, you will find a 3.3Ω resistor, and also 33Ω, 330Ω, 3.3kΩ, 33kΩ, 330kΩ and 3.3MΩ.
- Most resistors have their resistance value marked on them with a series of coloured bands. See Figure 6.1 for the colour code.

> **Exam tip**
>
> In the examination, you may be asked to use the E24 series of resistors. The E24 series contains the E12 series with 12 more values in between each one.

34   Now test yourself answers at www.hoddereducation.co.uk/myrevisionnotes

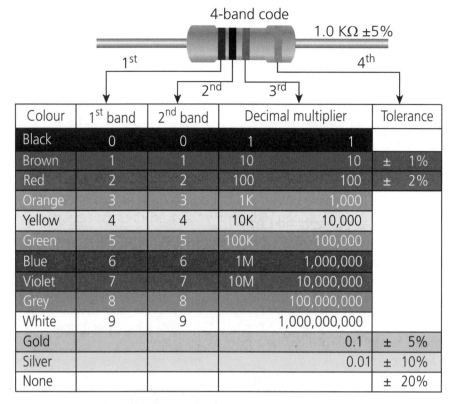

Figure 6.1 The 4-band resistor code

- Tolerance is measured in percent and it describes how close the resistor's actual resistance is compared to its printed value.
- For example, a resistor with band colours blue, grey, orange, brown has a resistance of:

  6 8 000 = 68 kΩ with a tolerance of ±1%.

Electronic system diagrams

- A system diagram clearly shows the input, process and output subsystems.
- It also shows the types of signals that pass between the subsystems.
- Signals can be analogue or digital.

Input devices

Switches

- A switch is a simple sensor capable of producing a digital input signal for an electronic system.
- **Latching** switches stay on when released while **momentary** switches turn off when released.
- A single-pole-single-throw (SPST) switch is the most basic on/off switch, with two terminals.

Typical mistake

Be sure to understand the differences between system diagrams, circuit diagrams and program flowcharts. They all show different types of information.

Latching: a switch which stays on when released.

Momentary: a switch which turns off when released.

- Single-pole-double-throw (SPDT) switches have three terminals, and they can be on in either position.
- Double-pole-double throw (DPDT) switches are two SPDT switches operated by a single lever.
- A **tilt switch** is on when it is upright and off when inverted. Tilt switches are used in electric heaters (to switch them off if they tip over) and motorbikes (to switch off the engine if the motorbike falls over).
- A **reed switch** is a SPST switch operated by a magnet. Reed switches are used as sensors in burglar alarms to indicate when a door is opened.

Analogue sensors

- Light-dependent resistors (LDRs) and thermistors were introduced in Topic 3.
- The resistance of an LDR decreases as light level increases.
- The resistance of a thermistor decreases as temperature increases (ntc type).
- These devices are connected in a **potential divider** to produce a voltage signal which changes with light level or temperature.
- If the LDR or thermistor is at the top of the potential divider, the output voltage rises as light or temperature increases.
- If the LDR or thermistor is at the bottom of the potential divider, the output voltage falls as light or temperature increases. See Figure 6.2.
- A moisture sensor can be made by replacing the LDR or thermistor with two parallel metal probes. When the probes are dipped in water, the output voltage from the potential divider will rise.

The potential divider formula is used to calculate the output voltage:

$$V_{out} = V_s \times \left(\frac{R2}{R1+R2}\right)$$

Process components

Transistor (MOSFET)

- A transistor is an **amplifier**. It turns a small signal into a bigger one.
- A transistor can be used as a **driver** to boost the signal from a process subsystem, so that it can switch on an output transducer.
- A MOSFET has three terminals, named drain (d), gate (g) and source (s).
- A MOSFET behaves like a switch connected between the drain and source terminals.
- The input terminal is the gate. The MOSFET switch will turn on when the gate voltage rises above 2.5 V.
- MOSFETs are very common components in electronic systems.

Thyristor

- A thyristor is used to switch on an output transducer when it receives a trigger signal.
- The thyristor stays switched on even when the trigger signal is removed. It is a latching device.

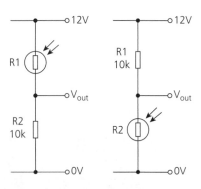

Figure 6.2 Two ways of wiring a potential divider

> **Potential divider**: two resistive components used to provide an analogue input signal for an electronic system.

> **Exam tip**
>
> Examination questions may ask you to identify an appropriate sensor for a given application and then draw a potential divider circuit to show how a voltage signal is produced. Practise drawing potential divider circuits for a range of sensing components.

> **Amplifier**: a subsystem to increase the amplitude of an analogue signal.
>
> **Driver**: a subsystem used to boost a signal so that it can operate an output device.

Figure 6.3 MOSFET circuit symbol

- The thyristor is reset by switching off the current through the transducer.
- The three terminals are named anode (a), cathode (k) and gate (g).
- The gate is the input trigger terminal.
- A thyristor is also called a silicon-controlled-rectifier (SCR).

Monostable

- A monostable is a timer circuit.
- When it is triggered, its output turns on for a period of time before turning off again.
- The time period is controlled by a resistor and a capacitor.
- A monostable can be built using a 555 **integrated circuit (IC)**.
- Monostables are used in security lights, burglar alarms and kitchen timers.

The time period for the monostable can be calculated using the formula:

$$T = 1.1RC$$

T is in seconds

R is the resistor value in ohms (Ω)

C is the capacitor value in farads (F)

Capacitors

- A **capacitor** is a component that stores electric charge.
- Capacitance is measured in units called farads (F), or microfarads (μF) or nanofarads (nF).

$$1\mu F = 1 \times 10^{-6} \text{ F}$$

$$1nF = 1 \times 10^{-3} \mu F = 1 \times 10^{-9} \text{ F}$$

- Capacitors below 1 μF are usually **ceramic** or **polyester** type.
- Capacitors over 1 μF are usually **electrolytic** type.
- Electrolytic capacitors are **polarised**.

Astable

- An astable circuit produces an output which repeatedly switches on and off. It produces a stream of output pulses.
- An astable can also be built using a 555 IC.
- The **frequency** is determined by two resistors and a capacitor.
- Astables are used in alarm circuits, clocks and flashing light systems.

The frequency can be calculated using the formula:

$$f = \frac{1.44}{(R1 + 2R2)C}$$

f is the frequency, in hertz (Hz)

R1 and R2 are in ohms (Ω)

C is in farads (F)

Figure 6.4 Thyristor circuit symbol

Integrated circuit (IC): a miniaturised, highly complex circuit with small pin connections in a single component.

Capacitor: a component to store charge, typically available in ceramic, polyester or electrolytic types.

Polarised: a component which has positive and negative leads and which must be connected the correct way around in a circuit.

Frequency: the number of pulses produced per second, in hertz (Hz).

Typical mistake

It is very easy to use the wrong units in monostable and astable formulae. Capacitance is often given in μF ($\times 10^{-6}$F) and resistance is often in kΩ ($\times 10^{3}\Omega$) or MΩ ($\times 10^{6}\Omega$). Practise, using your calculator, because many mistakes are made when typing in these values.

Voltage comparator

- A voltage comparator has two analogue voltage inputs, labelled V1 and V2.
- A voltage comparator compares the values of the two inputs.

  If $V_1 > V_2$ then V_{out} is logic 1 (on)

  If $V_2 > V_1$ then V_{out} is logic 0 (off)

- A voltage comparator is made from a component called an **operational amplifier** (op-amp).
- Voltage comparators are used in systems where it is necessary to switch on an output when the signal from an analogue sensor rises above or falls below a set level, for example, thermostats in heaters and refrigerators.

> **Operational amplifier (op-amp):** a specialised IC amplifier component.

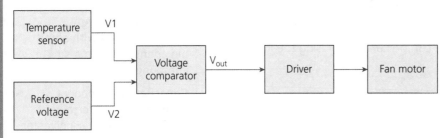

Figure 6.5 A voltage comparator system

Operational amplifiers

Voltage amplifier

- An amplifier turns a small signal (V_{in}) into a larger signal (V_{out}).
- The amplification factor is called the **voltage gain**.

$$\text{Voltage gain} = \frac{Vout}{Vin}$$

- A voltage amplifier can be made using an op-amp.
- Amplifiers are used in audio entertainment systems.

> **Voltage gain:** the amplification factor of an amplifier subsystem.

The gain of the voltage amplifier circuit is controlled by two resistors.

$$\text{Voltage gain} = 1 + \frac{Rf}{Ra}$$

Logic gates

- Logic gates are digital components.
- They process signals which are either logic 1 (high/on) or logic 0 (low/off).
- The output of a logic gate depends on the logic state of its inputs.
- You are required to know about six different gates: NOT, AND, OR, NAND, NOR and EOR.
- Each gate has a truth table which explains how the gate behaves. You need to learn the truth tables and the logic gate symbols shown in Figure 6.7.

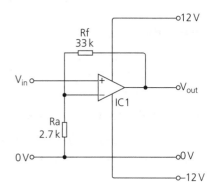

Figure 6.6 Voltage amplifier using an op-amp IC

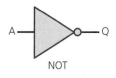

NOT

A	Q
0	1
1	0

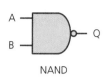

NAND

A	B	Q
0	0	1
0	1	1
1	0	1
1	1	0

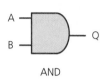

AND

A	B	Q
0	0	0
0	1	0
1	0	0
1	1	1

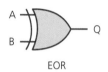

NOR

A	B	Q
0	0	1
0	1	0
1	0	0
1	1	0

OR

A	B	Q
0	0	0
0	1	1
1	0	1
1	1	1

EOR

A	B	Q
0	0	0
0	1	1
1	0	1
1	1	0

Figure 6.7 Logic gates and truth tables

Output components

Light output

- Light-emitting diodes (LEDs) are available in a variety of sizes, shapes and colours.
- Standard brightness LEDs are used as indicators, for example, an 'on' light on a product.
- High brightness LEDs are used in place of light bulbs for room illumination.
- LEDs have a positive terminal (the anode) and a negative terminal (the cathode).
- The cathode lead is usually the shorter lead, and it is often marked with a flat section on the case.
- A resistor must be placed in series with an LED to limit the current flowing.

The resistor value can be calculated using a modified version of the Ohm's law formula:

$$R = \frac{(V_s - V_{LED})}{I}$$

Where V_s is the power supply voltage, V_{LED} is the voltage drop across the LED and I is the current flowing through the LED.

Sound output

- A buzzer produces a tone when it receives power.
- A siren is a particularly loud or noticeable buzzer.
- A loudspeaker is used to produce a music or speech type of output sound. It cannot produce sound unless it receives a sound waveform.
- A piezo sounder is a miniature loudspeaker, often used to produce simple 'beep' tones.

Figure 6.8 An LED circuit symbol

Typical mistake

When drawing circuit diagrams, don't forget to draw a resistor in series with an LED.

Typical mistake

Practice using the formula to calculate the resistor value for an LED. A common mistake is to forget to work out $(V_s - V_{LED})$, and to just insert V_{LED} or V_s into the equation.

Typical mistake

Don't confuse buzzers with loudspeakers. A buzzer will only buzz. A loudspeaker will not produce any sound unless it receives a sound waveform, such as music.

Movement output

- Electric motors are very common components which produce rotary motion.
- A MOSFET (transistor) driver is needed to allow a process sub-system to control a motor.
- A **back emf** diode is needed when using a MOSFET to drive a motor (or a solenoid or relay). The diode protects the MOSFET by removing the back emf generated by the motor.
- A solenoid can provide a pulling or a pushing force when it receives power. It produces a short reciprocating motion.
- Solenoids are used in electric door locks and in valves for liquids or gases.

> **Back emf**: a high voltage spike produced when motors, solenoids or relays are used.

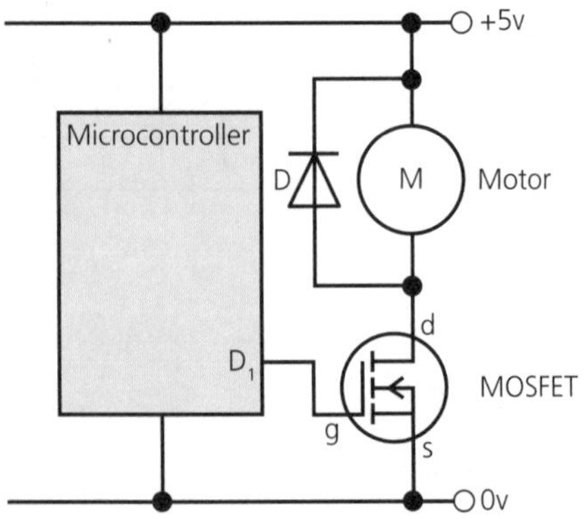

Figure 6.9 A MOSFET and back emf diode used to drive a motor

Relay

- A relay is a switch (called the 'contacts') which is controlled by an electromagnet (called the 'coil').
- Relays allow a high voltage (or a high current) output transducer to be controlled from a low voltage (or low current) circuit.
- Relay coils are often designed to work at 6 V or 12 V.
- Relay contacts are often SPDT or DPDT switches and these will have a maximum voltage and current rating, for example, 230 V, 10 A.

> **Exam tip**
>
> Don't forget to use a MOSFET and back-emf diode when drawing circuits showing motors, solenoids or relays as output devices. Practise drawing these circuits as the symbols are quite complex.

Functions of mechanical devices/systems

Rotary motion systems

- **Rotational velocity** is measured in revolutions per minute (rpm) or revolutions per second (rps).
- **Torque** is a turning force.
- In a simple mechanism, there is always a trade-off between rotational velocity and torque.
- A rotary mechanism can increase the torque but reduce the rotational velocity, or reduce the torque but increase the rotational velocity.

> **Rotational velocity**: the number of revolutions per minute (rpm) or per second (rps).
>
> **Torque**: a turning force.

Simple gear train

- A simple gear train consists of two interlocking spur gears. See Topic 4 for basic information on simple gear trains.
- The two gears rotate in opposite directions.
- The smaller gear rotates faster than the larger gear.

Now test yourself answers at www.hoddereducation.co.uk/myrevisionnotes

Velocity ratio

- The **velocity ratio** is the factor by which a mechanical system reduces the rotational velocity.

$$\text{Velocity ratio} = \frac{\text{rotational velocity of input}}{\text{rotational velocity of output}} = \frac{\text{number of teeth on output (driven) gear}}{\text{number of teeth on input (driven) gear}}$$

This equation can be rewritten as:

$$(\text{RV of input}) \times (\text{number of teeth on input})$$
$$= (\text{RV of output}) \times (\text{number of teeth on output})$$

Compound gear train

- Two or more sets of simple gear trains can work together to form a **compound gear train**.
- The overall velocity ratio is found by multiplying together the velocity ratios for each stage.
- Each stage of a compound gear train reverses the direction of motion, therefore, for a two-stage compound gear train, the input gear and output gear rotate in the same direction.
- Compound gear trains are used when it is necessary for a mechanical system to have a large velocity ratio.

For a two-stage compound gear train:

$$\text{Overall velocity ratio} = (\text{velocity ratio of stage 1}) \times (\text{velocity ratio of stage 2})$$

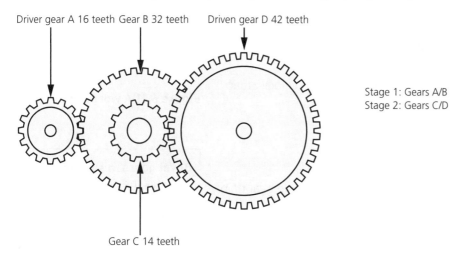

Driver gear A 16 teeth Gear B 32 teeth   Driven gear D 42 teeth

Stage 1: Gears A/B
Stage 2: Gears C/D

Gear C 14 teeth

Gears B and C are locked together

Figure 6.10 A compound gear train

Pulley and belt drive

A pulley and belt drive system differs from a simple gear train as follows:

- Pulley and belt systems transfer rotary motion between two shafts, which can be separated by some distance.
- The input and output pulleys rotate in the same direction.
- They are quieter during operation than a gear system.

The smaller pulley rotates faster than the larger pulley, in the same way as a simple gear train.

For the purposes of calculations, it is the diameters of the pulleys which determine the velocity ratio.

$$\text{Velocity ratio} = \frac{\text{diameter of output (driven) pulley}}{\text{diameter of input (driven) pulley}}$$

The key equation is written:

$$\text{(RV of input)} \times \text{(diameter of input pulley)}$$
$$= \text{(RV of output)} \times \text{(diameter of output pulley)}$$

Compound pulley system

- Compound pulley systems behave like compound gear trains, in that the overall velocity ratio is found by multiplying together the velocity ratio for each stage.
- In a compound pulley system, the input pulley and output pulley always rotate in the same direction, no matter how many stages there are.

Worm drive

- A **worm drive** consists of a worm screw (the input gear) and a worm wheel (the output gear).
- Worm drives achieve a very high velocity ratio.
- The velocity ratio is simply equal to the number of teeth on the worm wheel (for a single start worm screw).
- The direction of motion is transferred through 90°.
- They are self-locking, which means that the worm screw can drive the worm wheel, but not the other way around.
- Worm drives are used in winches and lifts where the high velocity ratio and the self-locking feature is particularly useful.

> **Worm drive**: a compact gear system which achieves a very high velocity ratio.

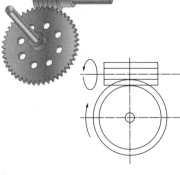

Figure 6.11 A worm drive

Bevel gears

- **Bevel gears** transfer rotary motion through 90°.
- The velocity ratio is calculated in the same way as for a simple gear train.

> **Bevel gears**: a system to transfer the direction of rotation through 90°.

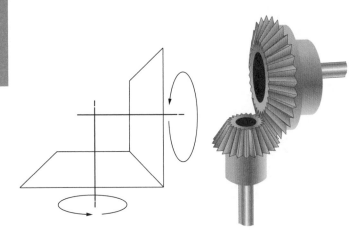

Figure 6.12 Bevel gears

Now test yourself answers at www.hoddereducation.co.uk/myrevisionnotes

Levers

A lever is a simple mechanism used to change forces and motion. See Topic 4 for basic information on levers.

- A lever can amplify force, or it can amplify distance moved, but it cannot do both at the same time. If a lever amplifies the force, then it will reduce the distance moved, and vice-versa.
- The **effort** is the input force and the **load** is the output force.
- The relative positions of the effort, load and **fulcrum** determine the class of lever.

In a **first-class lever**:

- the fulcrum is between the effort and load
- effort and load move in opposite directions
- the exact position of the fulcrum determines whether the lever amplifies the force, or amplifies the distance moved
- an example is a pair of scissors.

In a **second-class lever**:

- the load is between the effort and fulcrum
- effort and load move in the same direction
- the force is amplified (load is greater than effort), but the distance moved is reduced
- an example is a wheelbarrow.

In a **third-class lever**:

- the effort is between the load and the fulcrum
- effort and load move in the same direction
- the force is reduced (load is less than effort), but the distance moved is increased
- an example is a pair of tweezers.

Mechanical advantage

Mechanical advantage (MA) is the factor by which a system **increases** the force.

$$\text{Mechanical advantage} = \frac{\text{output force}}{\text{input force}}$$

For a simple lever:

$$\text{Mechanical advantage} = \frac{\text{input arm length}}{\text{output arm length}}$$

- If a system amplifies a force, its MA is greater than 1.
- If a system reduces a force, its MA is less than 1.

> **Mechanical advantage:** the factor by which a mechanical system increases the force.

Principle of moments

In a mechanical system:

Moment = force × perpendicular distance to fulcrum

The principle of moments states that, for a simple lever:

Effort × (input arm length) = load × (output arm length)

Ratchet and pawl

- Allows rotation in one direction only.
- Used in turnstile entry gates, in winches and lifts, and in ratchet spanners.

Ecological and social footprint

Changing society's view on waste

Topic 1 introduced the ways in which designers are gradually being forced to produce products which have a minimum impact on the earth as a result of:

- increasing consumer awareness, through the television and internet, of the problems caused by a throw-away society
- pressure from governments.

Living in a greener world

Electronic and mechanical products may contain:

- hundreds of different components
- a wide range of raw materials
- toxic materials such as lead, cadmium, mercury, sulphuric acid and radioactive substances.

If products are disposed of incorrectly:

- they can end up in a **landfill site**
- hazardous materials can leak out of them into the environment.

> **Landfill site**: a rubbish tip, where waste is simply buried.

The Waste Electrical and Electronic Equipment (WEEE) directive helps to reduce the damage caused by waste products by:

- making manufacturers and producers take responsibility for what happens to their products at the end of their lives
- requiring retailers to offer a free take-back service for old products, which must then be disposed of at an approved facility
- making a requirement for local councils to provide recycling facilities for electronic products.

Other issues related to recycling electronic products include:

- some materials that can be recovered are valuable, such as copper and gold
- designing the product to be easily separated into component materials
- accepting that redesigning products to be more recyclable can, initially, raise costs
- consumer pressure and market competition is likely to force designers and manufacturers to produce more environmentally-friendly products.

Sustainable design

Electronic products, such as mobile phones, can quickly become obsolete:

- software upgrades can cause the product to run slowly
- the battery can become worn out
- repairs can be costly
- the manufacturer stops product support
- there is a newer, more fashionable model on sale.

A **sustainable design** strategy is important for rapidly updated products:

- the manufacture of the product will be able to be sustained (kept going) with a minimal impact on the earth's environment or resources
- new materials will always be available
- useful materials will be recovered from discarded products
- a good business model will encourage consumers to return unwanted products and will create a circular economy where valuable resources are recovered from old products to be invested into new products (see Topic 1 for more on circular economy).

Exam tip

Exam questions may ask you to identify ways in which the manufacture and use of products can harm the environment, and ways in which designers can develop a sustainable design strategy.

Selection of materials and components

REVISED

A designer is faced with several factors to consider when selecting materials and components. These include:

- function – what does the component do in the system? This might also include parameters such as size, resistance and material
- aesthetic – is the look of the material important
- environmental – is the product exposed to rain, dirt, sunlight or extreme temperatures
- availability – does the supplier have sufficient stock for the batch to be manufactured
- cost
- social, cultural and ethical issues.

Figure 6.13 Filters in an air conditioning unit need to be regularly replaced

Components and their functional benefits or limitations

- The **rating** of a component is the maximum value of a specified quantity that it can handle.
- Operating a component beyond its rating is likely to damage it or reduce its life expectancy.
- Some components have a limited lifespan, such as batteries.
- Designers should consider how products can be designed for maintenance, so that servicing can be carried out to extend the product's life.

Rating: the maximum specified quantity a component is designed to handle.

Miniaturisation

- Prototype products made by hand are likely to be manufactured by different methods in larger scales of production.
- Industrially-made **printed circuit boards (PCBs)** are usually double-sided to enable complex circuits to be achieved. Some PCBs are multi-layered.
- Industrial methods use **surface mount technology (SMT)** to attach the components to the PCB. The components do not have wire leads.
- SMT relies on high speed pick-and-place robotic machines.
- The components are held in place by a solder paste, which turns into a permanent solder joint when the PCB is passed through a reflow oven.
- SMT allows complex, miniature circuits to be produced, which is essential for portable or wearable products.
- The PCBs are assembled at high speed, which reduces costs and increases the reliability of the end product.

> **Printed circuit board (PCB)**: a board with a pattern of copper tracks which complete the required circuit when components are soldered on.
>
> **Surface mount technology (SMT)**: the industrial method of mounting miniature components onto a PCB using robotic machines.

Cultural, social, ethical and environmental responsibilities of designers

- The Restriction of Hazardous Substances (RoHS) directive reduces the use of materials such as lead, cadmium and mercury in electronic equipment.
- Hazardous substances can be found in batteries, paints, solder and polymers.
- Designers have an ethical responsibility for the working conditions of the people who manufacture the products.
- Designers have a responsibility to ensure that the materials used in the products are ethically sourced.
- The Ethical Trading Initiative (ETI) promotes workers' rights around the globe.

The impact of forces and stresses

REVISED

How mechanical components are strengthened to withstand forces

- A **structure** is a collection of parts that work together to provide support.
- **Members** are the individual parts in a structure.
- The base supporting structure is the **chassis**.
- A **rigid** system can withstand forces without bending.

When a force is applied to a component, the amount it bends or flexes depends on:

- the size of the force
- the material
- the thickness of the component.

It can also depend on:

- the cross-sectional shape of the component
- where the force is applied in relation to the supports.

> **Structure**: a collection of parts to provide support.
>
> **Member**: an individual part in a structure.
>
> **Chassis**: the base supporting structure.
>
> **Rigid**: will withstand forces without bending.

Changing the cross-sectional shape can increase rigidity while reducing the weight and cost of the component.

- Large sheet materials are made more rigid by adding folds or indents.
- **Ribs** are thin supports attached to sheet materials to increase their rigidity.
- **Triangulation** is a method of increasing the rigidity of a structure by designing it to be composed of triangles rather than rectangles.
- A triangle is a naturally rigid shape. A rectangle is an unstable shape which relies on its corner joints to keep it square.

> **Ribs**: thin supports applied at right angles to a sheet to increase its rigidity.
>
> **Triangulation**: the method of increasing rigidity by designing a structure to compose of triangles.

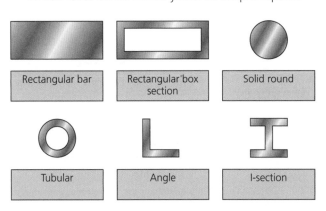

Figure 6.14 **Examples of material cross-sections**

Figure 6.15 **Waterproof camera case**

Casing and protecting electronic components

In an electronic system, a casing may be used for:

- protecting the components from the environment
- providing structural support, for example, holding the product together
- aesthetic reasons.

Water entry inside a case can damage the components:

- products can be made splash proof using simple covers over key parts, and by careful design of the layout of sensitive parts
- fully waterproofing a case is difficult. Rubber seals can be used on sections of the case which need to open, for example, a battery compartment
- waterproofing is compromised by switches, sockets and other holes

> **Exam tip**
>
> Exam questions may ask you to study a photograph of a structure and identify how it has been designed to withstand forces and stresses without failing.

Stock forms, types and sizes

REVISED

Standard stock electronic component sizes

- Most electronic components for prototyping have leads which will fit through circuit board holes on a 2.54 mm (0.1 inch) grid.
- Many components including resistors and capacitors are manufactured in certain values, called preferred values.

Dual-in-line standard for electronic ICs

- **Dual-in-line (DIL)** ICs have pins which fit the standard 2.54 mm (0.1 inch) spacing.

> **Typical mistake**
>
> It is easy to forget that the pins are numbered anticlockwise around a DIL IC. The highest and lowest pin numbers are always opposite each other.

> **Dual-in-line (DIL)**: a standard package style for ICs.

- There is a notch in one end of the IC, and pin 1 is to the left of this. Pin 1 is sometimes identified by a small dot.
- The pins are numbered anticlockwise around the case.

DIL sockets are often used for prototype PCBs.

- They prevent the IC being accidentally overheated when soldering.
- The IC can be removed or replaced if it becomes faulty.
- The IC can be inserted at the last stage of assembly, meaning it is less likely to be damaged by static electricity, which can happen if the IC is handled excessively.
- DIL sockets slightly increase costs, and they increase the dimensions of the PCB.

Stock materials for the manufacture of products

Stock forms are standard shapes and sizes of materials.

Woods

- A softwood timber, such as Scots pine, is often used to provide structural support to an engineered product.
- Planed all round (PAR) timber has a smooth surface, and the finished size is about 3 mm less on each surface than the quoted size.
- Natural timber will vary between batches and may twist or warp with the weather.
- Manufactured boards such as plywood or medium-density fibreboard (MDF) are more stable than timber and available in large sheets.

Metals

- Steel is a ferrous metal and it will rust if it is not protected.
- Aluminium and brass are non-ferrous, non-magnetic and do not require a finish to be used outdoors. Aluminium has a good strength-to-weight ratio and brass has a pleasing aesthetic finish.
- Metals are available in sheets, and as long bars with various profiles, for example, pipe, box section or angle.

Polymers

- Thermoforming polymers soften when heated and can be formed into shape.
- Standard stock forms are sheets, extruded sections, reels, powders or granules.

Manufacturing to different scales of production

REVISED

One-off (bespoke) production

- Expensive because the designer will not receive rewards from any future sales.
- Often hand-made, requiring a high level of skill.
- **Rapid prototyping** may be used (see below).

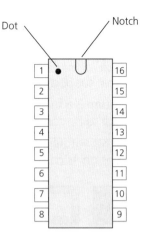

Figure 6.16 DIL pin numbering

> **Stock forms**: the standard shapes and sizes of materials that are commonly available.

> **Rapid prototype**: an accurate, finished 3D, part produced from a CAD model in a matter of minutes or hours.

Batch production

- A method of manufacturing a pre-determined number of items.
- There is no limit on batch size, but production will continue until the entire batch is completed.
- Manufacturers organise a production system to suit their machinery and their workforce.

Mass production

- Used for the manufacture of commonly used items, such as screws and batteries.
- Specialist factories produce items quickly at high speed.
- Repeatability and consistency of the items is achieved.
- In **continuous flow production**, items are produced 24 hours a day.
- The cost per item is usually extremely low.

> **Continuous flow production**: a non-stop production method which produces high volume parts at low cost.

Just in time (JIT) manufacturing

- Achieves manufacturing efficiency by ensuring that materials are ordered to arrive just before they are needed for a batch manufacture.
- The finished products are shipped out soon after completion, so there are no storage costs.
- JIT manufacturing achieves efficient flow through a factory.

The use of CAD/CAM in production

- 2D and 3D **CAD** software can link directly to **CAM** machines such as a laser cutter, CNC router, plasma cutter, vinyl cutter or CNC lathe.
- CAM machines achieve accurate and repeatable cuts in materials.
- A 3D CAD design can be quickly realised on a 3D printer. This is called rapid prototyping.
- PCB design software is used to route the PCB tracks in complex circuit designs and to test the function of the design.

> **CAD/CAM**: computer-aided design/computer-aided manufacture.

Jigs and devices to control repeat activities

During manufacturing, it is important to ensure that repetitive tasks are done in a way that achieves repeatability. Specialist equipment may be made to guide workers and their tools:

- A jig guides a tool so that it cuts in the right place.
- A fixture holds the material in exactly the right place so it can be processed accurately.
- A template is used for repeated marking out onto materials.
- A former is used to achieve a precise bend angle.

Specialist techniques and processes

REVISED

Wastage/addition

You should be familiar with the application and use of the following tools.

- Marking out tools including steel rule, try-square and marker/scriber.
- For holding work during fabrication, a bench hook, vice and G-clamp.

- Hand cutting tools including a tenon saw, coping saw, hacksaw, tin snips and nibbler for sheet metal.
- Machine cutting tools including a scroll saw and bandsaw.
- Shaping tools including a file, needle file and sanding machine.
- Drilling tools including a cordless drill, bench drill, twist drill, Forstner bit, flat bit, hole cutter and cone cutter.

Deforming/reforming

You should be familiar with the following methods.

- Bending polymers – line bending, drape forming.
- Vacuum forming – shape and design of former, draft angle, vent holes.

Hot/cold working of metals

You should be familiar with the following methods.

- Cold working – folding, punching, rolling. Cold working can cause metal to become hard and brittle.
- Hot working – **annealing** to increase **ductility** to prevent cracking when bending.
- Casting – industrial casting commonly uses iron, brass or aluminium. Casting in school usually uses pewter, an alloy which melts around 230°C.

> **Annealing**: a heat treatment to reduce cracking when bending metal.
>
> **Ductility**: the property of a material to be able to be permanently stretched out without cracking.

Other processes

Turning:

- using a centre lathe for metal
- using a wood lathe.

Laser cutting:

- setting laser power and speed of cut
- focusing the laser
- need for material to be flat.

3D printing:

- materials used include polymers, metals, ceramics or food
- fast production of prototype parts
- useful for bespoke production.

Assembly and components

Temporary joining methods

Nuts and bolts:

- requires a clearance hole to be drilled through both parts
- bolt identification labelling, for example, M4 20.

Self-tapping screws (for metal) and wood screws:

- requires a clearance hole through one part and pilot hole into the other part.

Permanent joining methods

Pop rivets for sheet materials:

- requires a clearance hole through both parts.

Adhesives:

- need for choice of correct adhesive depending on the materials to be joined
- need for cleaning and preparation of surfaces
- types of adhesive and their application: PVA, contact adhesive, Tensol, epoxy resin, hot melt.

Soldering:

- for making permanent electronic joints between components and on circuit boards
- solder is an alloy which melts around 220°C
- use of flux to clean the joint.

Brazing:

- carried out at a higher temperature than soldering
- can be used to join steel, aluminium, copper and brass, for example
- a filler metal is used in the join.

Welding:

- the strongest method of joining metals
- the metals melt and fuse together.

Exam tip

Questions relating to manufacturing will score the highest marks if you use the precise names of tools, processes and materials rather than generic words such as 'saw', 'moulding' or 'plastic'.

Surface treatment and finishes

REVISED

Surface finishes applied to electronic devices

Casings are used to protect electronic systems and to improve their aesthetic appearance.

Polymers

- Casings made from polymers are likely to be **self-finished**, so they do not require any additional surface finishes.

Metals

- Steel will rust unless it is protected with a finish. Finishes for steel include a film of oil or paint.
- Before applying paint, the surface must be cleaned and a coat of **primer** applied before the final coat(s) of paint in the desired colour. Paint can be applied by brush or spray.
- Aluminium and brass do not rust, but they do oxidise, so they are often polished and a clear lacquer is applied to maintain an attractive look.
- Brushed and lacquered aluminium is a popular finish.

Self-finished: a material that does not require the application of a finish to protect it or improve its appearance.

Primer: the base coat of paint applied straight to the material surface.

Woods

- A **preservative** can be used to prevent wood decaying when used outdoors.
- Wood can be painted. The surface is first sanded, then a primer coat is applied. Further coats are then applied, lightly sanding between each coat.

Powder and polymer coating of metals

- Powder-coating is a form of painting.
- The metal is cleaned **by shot-blasting**.
- A dry, coloured polyurethane powder is sprayed onto the metal, using an electrostatic charge to encourage the powder to stick.
- The metal is baked in an oven, causing the polymer powder to melt and fuse into a smooth coating.
- Powder coating is used on items such as fridges, washing machines and bicycle frames.

Dip-coating is a similar process:

- the metal part is heated
- it is then plunged into a **fluidised bath** of coloured polymer powder
- the powder melts and fuses to the metal.
- Dip-coating is used on items such as tool handles, cupboard door handles and coat hooks.

> **Preservative**: a chemical treatment applied to wood to prevent biological decay.
>
> **Shot-blasting**: using grit, fired at high pressure, to clean a surface by abrasion.
>
> **Fluidised bath**: blowing air through a powder to cause it to behave like a fluid.

Figure 6.17 Powder coating an alloy wheel

Now test yourself

TESTED

1 (a) Draw a system diagram, based around a voltage comparator, for a night light which turns on an LED when the light level drops below a threshold. [4]
  (b) Draw the circuit diagram for the night light from part (a). [4]
  (c) Calculate the LED resistor value if the high output from the voltage comparator is 12 V, the LED has a voltage drop of 2.2 V and the LED current is 8 mA. [3]
  (d) Select the nearest preferred resistor value from the E12 series. [1]
2 Draw the circuit symbols and explain the action of the following electronic components:
  (a) Thyristor [2]
  (b) MOSFET [2]
  (c) Relay [2]
3 Five spur gears are available with the following number of teeth:

   15t   20t   30t   40t   60t

   Draw a diagram to show how four of these gears can be used to make a compound gear train with a velocity ratio of 6. [4]
4 Draw three labelled diagrams to show the differences between the three classes of lever. [3]
5 Describe one way in which legislation has impacted on how society disposes of unwanted or obsolete products. [3]
6 Describe the benefits offered by the use of surface mount technology (SMT) in electronic products. [3]
7 Describe, with examples, two reasons why electronic systems and components are often placed in a casing. [4]
8 Metal bars are available in a variety of stock profiles. Sketch and name three different metal bar stock profiles. [3]
9 Give two reasons why one-off production is expensive. [2]
10 Use sketches and notes to describe the steps involved in casting a pewter circle, about 40 mm diameter. [8]
11 Describe the differences between brazing and welding as methods of permanently joining metal plates. [4]
12 Describe the powder coating process for painting steel wheels. [4]

7 Papers and boards

Sources, origins, physical and working properties REVISED ▢

Paper

- Paper was originally made from fibres of tree bark mixed in water, known as **pulp**. The pulp was drained, spread out and pressed into a thin layer before being dried.
- It was later discovered that lignin, the natural glue that holds the fibres together, could be broken down more easily if plants with long cellulose fibres were used. These fibres could be made into a finer pulp which made better quality paper.

> **Pulp**: raw material from trees used to make paper.

Making paper by hand

- Rip up old used paper into small pieces (approx. 2 cm square) and soak in water.
- Gradually add to water in a blender until the mixture has become a pulp that looks like wallpaper paste then pour into a tray.
- Spread out the pulp evenly then flatten and squeeze out as much moisture as possible.

The mechanical pulping process

- Paper is made in a similar way to the original method, but mechanical methods are used to separate the wood fibres. See Figure 7.1.
- Mechanically pulping paper damages the fibres more than by hand.
- The paper has a low strength, tears easily and disintegrates when wet.
- Mechanically pulped paper is suitable for paper products that use 'bulk' grades of paper such as newspaper and toilet tissue.

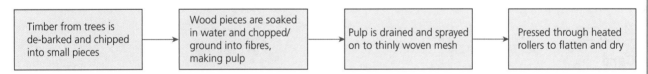

Figure 7.1 The mechanical pulping process

The chemical pulping process

- The chemical process uses chemicals to break down the wood pieces and removes the lignin without damaging the fibres, creating much stronger pulp and higher quality paper.
- Bleaching agents, dyes and fillers are added to give the paper a specific colour or property.

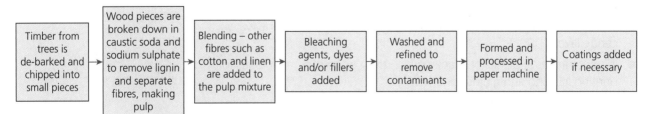

Figure 7.2 The chemical pulping process

Recycled paper

- Recycled paper reduces the number of trees needed to make paper and lessons the environmental impact.
- Paper cannot be recycled indefinitely as after around five or six times, the fibres become too short and weak to pulp adequately.
- To maintain the quality and strength of recycled paper, a mixture of recycled paper and new virgin wood chippings are used to make the pulp (see Table 7.1).

Table 7.1 Amount of recycled paper used in making paper

Pulping process	% recycled paper	% virgin
Mechanical pulping	55%	45%
Chemical Pulping	77%	23%

Gsm: grams per square metre – used to measure the weight of paper.

Table 7.2 Physical and working properties of paper

Common name	Weight (gsm)	Properties/working characteristics	Uses
Layout paper	50	Bright white, smooth, light weight (thin) so slightly transparent and inexpensive	Sketching and developing design ideas. Tracing parts of designs
Copier paper	80	Bright white, smooth, medium weight, widely available	Printing and photocopying
Cartridge paper	80–140	Textured surface with creamy colour	Drawing with pencil, crayons, pastels, watercolour paints, inks and gouache
Bleed proof paper	80–140	Bright white, smooth surface, stops marker 'bleed'	Drawing with marker pens
Sugar paper	100	Available in wide range of colours, inexpensive, rough surface	Mounting and display work

Boards

Card

- One method of making cardboard is to sandwich and paste multiple layers of paper together.
- Another is to press the layers of wet pulp together into a thicker layer.

Corrugated card

- Corrugated cardboard has three layers made by passing paper through a corrugation machine.
- The centre layer is steamed to soften the fibres, then crimped to give it a wavy shape.
- The two outer layers of paper are then glued on each side of the wavy centre layer.
- It is cut into large pieces or 'blanks' which then go to other machines for printing, cutting and gluing together.
- Double walled corrugated card has an additional wavy and flat layer to make it more rigid and give extra protection.

Now test yourself answers at www.hoddereducation.co.uk/myrevisionnotes

Board sheets

- Mounting board is a smooth, rigid type of card, around 1.4 mm (1400 **microns**) thick.
- It is available in different colours but white and black are the most common.

Table 7.3 Physical and working properties of card

Common name	Thickness (microns)	Properties/working characteristics	Uses
Card	180–300	Available in a wide range of colours, sizes and finishes, easy to fold, cut and print onto	Greetings cards, paperback book covers, as well as simple modelling.
Cardboard	300 upwards	Available in a wide range of sizes and finishes, easy to fold, cut and print onto	General retail packaging such as food and toys. Design modelling
Corrugated cardboard	3,000 upwards	Lightweight yet strong, difficult to fold, good heat insulator	Pizza boxes, shoe boxes, larger product packaging such as for electrical goods
Mounting board	1,400	Smooth, **rigid**, good fade resistance	Borders and mounts for picture frames

Laminated layers

Foam board and Styrofoam

- Foam board and Styrofoam™ are slightly different types of polystyrene foam.
- Foam board uses this foam sandwiched between two outer layers of paper or thin card.
- Styrofoam is a trade name for **extruded** polystyrene foam insulation (XPS). XPS is made by melting plastic resin and other ingredients into liquid form and extruding through a die.
- The extruded liquid then expands as it cools producing a closed cell rigid insulation.

For more information on expanded polystyrene foam see Topic 10.

Corriflute

- Corriflute is a trade name for corrugated polypropylene plastic.
- It is manufactured in one piece by extruding molten polypropylene through a former that moulds the polymer into the shape required.

> **Micron**: one thousandth of a millimetre (0.001 mm) – used to specify the thickness of board.
>
> **Rigid**: inflexible, stiff.
>
> **Extrusion**: drawing material through a shaped former to create the shape required.

> **Exam tip**
>
> Corriflute, PVC foam and Styrofoam are made from polymers and therefore have the same basic characteristics of polymers. For example, they are waterproof and deform with heat.

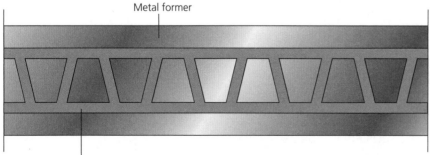

Metal former

Polymer is drawn through voids in the former to form the required shapes

Figure 7.3 Section through Corriflute

Foamex

- Foamex is a tradename for polyvinylchloride (PVC) foam.
- PVC foam is made by mixing two chemicals (diisocyanates and polyols) which react with each other.
- Pigments for colour and other additives are added.
- The mixture is then poured onto a moving conveyor belt where it 'foams', forming one long continuous block of foam.
- It is available in a range of different colours and is easy to cut and join to other materials.

For information on PVC please see Topic 10.

Table 7.4 Properties of types of board

Common name	Properties/working characteristics	Uses
Foam board	Smooth, rigid, very lightweight and easy to cut	Point-of-sale displays, ceiling-hung signs in supermarkets, architectural modelling
Styrofoam	Light blue colour, easy to cut, sand and shape, water resistant, good heat and sound insulator	3D moulds for vacuum forming and GRP, wall insulation in caravans, boats etc.
Corriflute	Wide range of colours, waterproof, easy to cut, rigid, lightweight	Outdoor signs, packaging and modelling
Foamex	Lightweight, good insulation properties, easy to print on, water resistant	Outside displays and signs

Ecological and social footprint

Paper and card

- Seventeen trees, thousands of gallons of water and kilowatts of energy are needed to make one tonne of paper.
- Recycling paper means natural resources, energy and emissions can be drastically reduced along with landfill space.
- Recycled paper also produces 73 per cent less air pollution than if it was made from raw materials.
- The **Forest Stewardship Council (FSC)** run **managed forests** where one or more trees are re-planted for every one felled.
- Bleaching agents, dyes and additives are also used in the manufacturing process. These can be harmful to the environment and our health, causing irritation to the eyes, and skin and breathing problems.
- The lignin extracted during the pulping process can be burnt as a fuel oil.

> **Forest Stewardship Council (FSC)**: an organisation that promotes environmentally appropriate, socially beneficial and economically viable management of the world's forests.
>
> **Managed forest**: a forest where new trees are planted whenever one is cut down.

Laminated boards

- Laminated boards are much harder to recycle.
- The polymers need to be separated, sorted, cleaned and chipped before being melted down into useable granules.
- This is a time and energy consuming process, meaning many laminated boards end up in landfill.

The ecological, social and ethical issues associated with processing the plastic elements in laminated layers are covered in Topic 10.

Life cycle

- Paper and cardboard decompose quickly compared to other materials – two to four weeks for paper and around two months for cardboard.
- Paper and cardboard do not release any harmful chemicals during decomposition.
- Laminated layers take much longer to decompose as they contain polymers. Styrofoam can take around 50 years to decompose and PVC can take up to 500 years.
- Laminated layers that contain Styrofoam or PVC release toxic chemicals.
- Approximately 70 per cent of paper and card products are recycled. Paper fibres that cannot be recycled any further are left to decompose or burned as a fuel source.
- A small percentage of paper and card cannot be recycled and ends up in landfill. For example, food packaging is often not recycled due to being contaminated by grease, oil and other liquids. These are absorbed into the paper fibres and create dark marks and holes that make the paper unsuitable for any use.
- Food-soaked paper and cardboard can be composted. Composting is a natural process that benefits the environment by creating new soil that is rich in minerals.

Figure 7.4 Used food box showing contamination with grease

> **Life cycle**: the stages a product goes through from beginning (extraction of raw materials) to end (disposal).

> **Exam tip**
>
> Papers and cardboard are made from trees and therefore products made from them have a similar life cycle to timber products.

Factors affecting the selection of materials and components

REVISED ☐

Properties of the material

The main influence on selection of material will be the properties and characteristics of the material. The following properties need to be considered:

- surface finishes – these can affect strength and rigidity and can have a gloss (shiny) or matt finish
- absorbency – the lower the absorbency, the higher quality of the image when printed onto it
- colour – many products require a specific colour paper to be used
- texture – the rougher the texture the better the material is to draw or paint on
- flexibility/rigidity – some products need to bend whereas others require a rigid, stiff material
- water resistance – certain outdoor applications require waterproof boards such as Corriflute.

Type of printing process

- Modern printing methods can print onto a wide range of materials and different shaped products.
- Traditional printing methods cannot print onto materials such as corrugated card, foamboard or Corriflute.

Cost

- Low manufacturing costs maximise profit, making the product as cheap to buy as possible.

Scales of production

- One-off products are usually hand-made by craftsmen.
- Batch-produced products use a small team with different skills who carry out a specific part in the manufacturing process.
- Mass-produced products use production lines with human and/or automated assembly.
- For more details, see 'Manufacturing to different scales of production' below.

Availability

- Products may require sizes, colours or textures that are non-standard and have to be made to order.
- These take longer to produce, are made in limited quantities and are more expensive.

Environmental influences

- **Deforestation** – many paper manufacturers choose to use only trees from managed forests such as those run by the FSC.
- **Water and energy consumption**:
  - paper production can lower the water table due to the amount of water required, and increase water temperature and sedimentation which affects local wildlife
  - paper-making machines also use large amounts of electricity
  - many paper mills now recycle up to 90 per cent of the water used.
- **Chemical and solvents** – unbleached paper can be used for products that do not need to be white by reducing toxic solvents and chlorine.
- **Air pollution** – pulp and paper mills produce carbon dioxide and other pollutants that damage the ozone layer, cause acid rain and contribute to global warming.
- **Solid waste** – waste paper fibres form a sludge which is either disposed of into landfill or can be dried out and burned as a fuel.

> **Deforestation**: the large-scale felling of trees which are not replanted.

Social, cultural and ethical issues

- Imagery, symbols and even certain colours can mean different things in another **culture**.
- Designers must take care how they portray individuals or groups of people and when selecting and using photographs or images of people on their products. Consideration must be given of how an image may represent a minority group. Using a person of a particular race can sometimes portray them in a negative light or stereotypical manner, which can be extremely offensive.
- Designers must consider different social attitudes and how these differ across the world. What is socially acceptable in the West may not be in other countries.
- In other cultures there are different beliefs and attitudes regarding revealing clothing and keeping certain parts of the body covered at all times.

> **Culture**: the ideas, customs and social behaviours of a particular people or society.

Responsibilities of designers

- Due to globalisation, many products are produced in developing countries where labour rates and materials are much cheaper.
- In the West, health and safety laws protect the safety, rights and welfare of all workers. Developing countries have less stringent safety laws and poor people can be exploited.

Exploitation is when workers are forced to work:
  o in unsafe, unhealthy or dangerous conditions
  o for extremely long hours without sufficient breaks
  o without correct protective equipment
  o for low pay rates that do not reflect a fair wage.

- Designers have a responsibility to 'refuse' to design products for companies who exploit their workers in this way.
- Organisations such as Fairtrade ensure that workers are not exploited.
- Consumers can support the rights of workers by refusing to purchase goods not approved by these organisations.
- Designers have a responsibility to design products that are as eco-friendly as possible.
- Using recycled paper and card whenever possible and designing products that can be easily recycled is one way of doing this, for example, avoiding the use of coatings that make recycling paper and card products difficult.

> **Exploitation**: treating someone unfairly in order to benefit from their work.

Estimating the true costs of a prototype or product

- The price of producing a prototype or final product will vary depending on many factors such as:
  o the quantity required
  o the size and complexity of the product
  o the materials and processes used to make it
  o the production time
  o the price and availability of any manufactured components.
- Designers should carefully consider and calculate the amount of material needed.
- The costs of paper and boards reduce the more you purchase.
- The cost per mm² of all sheet materials decreases as the sheet size increases.
- It is more economical to buy a large size sheet and cut it down into smaller sheets than buying them ready cut.
- Large sheets are more difficult to transport so the designer must also factor in transport and delivery costs.
- Buying materials from countries on the other side of the world can often be cheaper than buying in your own country.
- The effects of globalisation and local economies must be considered when deciding where to purchase products.

The impact of forces and stresses

REVISED

- Structural integrity is a product's ability to hold together under a load without bending or breaking.
- Structural failure occurs when an object does not hold together because it has been stressed more than it can withstand.
- Paper can appear to have little structural integrity, but it can be greatly increased in various ways.

Folding

- Folding a sheet of paper turns it from a flat sheet into a structure.
- The fold creates a rigid section that can support its own weight and additional weight.
- Folding the paper a number of times further increases its structural integrity.

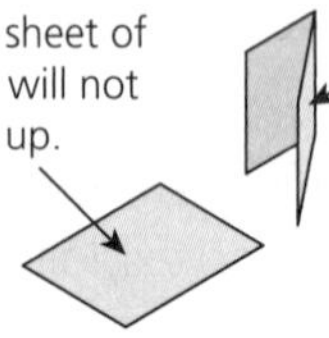

A flat sheet of paper will not stand up.

Once folded the sheet of paper can stand up.

Figure 7.5 Effect of folding paper

Curving and bending paper and card

- Paper and card can be bent into curved shapes which have excellent strength and structural stability.
- The wavy centre section of corrugated card gives it rigidity and strength.
- The triangulated shapes in the centre of Corriflute work in the same way.

Stiffening papers and boards

Thin card and paper can be stiffened in a number of ways:

- by sticking together several sheets to increase thickness
- by adding ribs to give extra strength in one direction.
- by laminating (encasing the paper in thin sheet plastic).

Monocoque: structural system where loads are supported through an object's external skin.

Reinforcement: extra material added to increase strength.

Structural integrity of card

- The structural integrity of cardboard can be increased by shaping it into a structure, for example, cardboard egg boxes, which are known as a '**monocoque**' structure.
- Structural integrity can also be increased by **reinforcing** one material with another. An example is foam board where the inside of the board allows the outer paper layers to stand on their edge and not bend as easily. The paper outer layers spread any load across the foam instead of it being concentrated in one place.

Typical mistake

Paper and cardboard are often considered to not be strong materials but when folded or shaped in certain ways to create 'structures' they are incredibly strong for their weight.

Stock forms, types and sizes

REVISED

Sizes

- Paper and board sizes range from A10 through to 4A0 – see Figure 5.1 in Topic 5. The most common sizes used by designers are between A6 and A0.
- Foam board is available in standard sizes from A4 to A0 and thicknesses of 3 mm, 5 mm or 10 mm. Many suppliers stock standard imperial sizes up to 8 ft x 4 ft (2440 mm x 1220 mm).

- Corriflute is available in a range of standard sizes. It is available in thicknesses of 2 mm to 10 mm and comes in a range of different colours.
- PVC foam is available in standard paper sizes and thicknesses of 1, 2, 3, 4, 5, 6, 8, 10, 13, 15, 19 and 25 mm. A range of standard and special designer colours with gloss or matt finishes are available.
- Styrofoam is available in sheet form or in blocks. Sheet thicknesses range from 5 mm to 165 mm through increments of 5 mm or 10 mm. Thicknesses above 165 mm are considered a block rather than a sheet.

Costs

- Paper and card products can be costed by multiplying the cost per sheet by the number of sheets required.
- Paper is much cheaper per sheet when bought in **reams**.
- The size of card required can be found by finding the optimum sheet size for the number of items required (see tessellation).
- Corriflute, foam board and Styrofoam costs can be calculated by working out the combined surface area of all the pieces required and the minimum standard sheet size they will fit onto.

Ream: pack of 500 sheets.

Manufacturing to different scales of production

REVISED

One-off production

- **One-off production** is the making of a single product.
- One-off products are labour intensive, time consuming and the most expensive way of producing an item.
- Some one-off products are entirely hand-made whereas others can use automated processes.
- One-off products are often produced for a specific client's requirements or needs.
- A hand-made card is an example of one-off production.

One-off production: process used when making a prototype product.

Batch production: making a small number of the same or similar product.

Batch production

- **Batch production** is where a limited number of an item are produced in one go.
- Similar products with slight variations such as changes to text or colour can also be produced.
- For printed products, digital printing is the most cost-effective batch production method. Digital printers are inexpensive to buy, readily available and many people have them for household use.
- The cost per sheet of printing using a digital printer is high compared to other types of printing but they are the best option for small print runs.
- Screen printing is another method of printing suitable for batch production.
- Screen printing is used for creating repeating patterns or designs. The lightest colour is printed first then the screen is washed and masked up for the next colour. The process is repeated until all the colours have been printed.

Stencils and templates

- Stencils and templates can be used for batch production of paper and board products.
- The template is laid onto the material and drawn around, then moved and drawn around again until the required number is reached.
- A template ensures that every piece is exactly the same and saves the time of drawing each part out individually.
- **Tessellation** reduces waste by arranging the pieces in such a way that the material being cut out is used to its maximum capacity.

Computer-Aided Manufacturing (CAM)

Vinyl cutting

- Most schools have a vinyl cutter available of at least A4 size.
- The most common uses are for creating logos, designs and text to stick onto packaging and point-of-sale displays where they cannot be printed.
- Vinyl-cut lettering and images are often used on the sides of cars, buses and lorries.
- Vinyl graphics can be removed relatively easily by applying a little heat to soften the vinyl and peeling them off.

> **Tessellation**: an arrangement of shapes closely fitted together in a repeated pattern without gaps or overlapping.

Large-scale (mass) production

Offset lithography

- Offset lithography is one of the most common forms of commercial printing. It uses a **colour separation** printing process.
- Four ink colours: cyan, magenta, yellow and black (shortened to CMYK) are used.
- The colours are overlaid to create others, for example, cyan on top of yellow creates green.

Flexography

- Flexography is another type of mass production printing process.
- The process is quicker and cheaper than lithography, although the quality is not as good.
- It is used for printing on packaging where the quality of print is not as important.

> **CAM**: computer-aided manufacturing.
>
> **Mass production**: making products in very large numbers.
>
> **Colour separation**: separate colours (cyan, magenta, yellow and black) printed in different combinations to create other colours.

Pre-press

- Once a design has been drawn on paper or produced on a software program, it must go through the pre-press stages before it is ready to be printed.
- The first stage checks the fonts and formatting are correct.
- The resolution (sharpness) and colours of images are checked to ensure the four separately printed CMYK colours produce the required colour when printed.
- The layout of the page is then checked to ensure it fits on the page and in the correct position.
- Registration marks printed on the edge of the page are used to line up the different plates for multi-colour printing jobs.

Now test yourself answers at www.hoddereducation.co.uk/myrevisionnotes

- The imposition checks the arrangement of pages on the printer's sheet are optimised to allow faster printing, simplify binding and reduce paper waste.
- Once all the checks have been made, a 'proof' of the document is created and sent to the client for final approval.

Binding methods

- Different binding methods are used for books, magazines and other large publications.
- Most publications are sewn together.
- Saddle stitching is the most common and least expensive method used for books and magazines.
- Loop stitching works in a similar way but allows extra pages to be added at a later date.
- Side or stab stitching uses wire then a section along the edge is added to cover the wires.
- Sewn binding is similar to saddle stitching but uses thread instead of wire.
- Documents with large numbers of pages are bound in stages. The pages are divided up and stitched into small sections called signatures. The signatures are then joined together using one of the methods below:
  - perfect binding is where the signatures are glued together into a wrap-around cover
  - tape binding is similar to perfect binding but uses an adhesive tape wrapped around the signatures to hold them all in place
  - case binding involves gluing the signatures to end papers which are glued to the spine of the book cover.
- Other binding methods include:
  - ring or comb binding – uses holes punched along the edges of the pages and a spiral ring or plastic comb binder holds the pages loosely together
  - plastic spines – a u-shaped length of plastic that grips the pages placed between them
  - stud binding (also known as screw or post binding) – holes are drilled through the pages, then a stud is pushed through and an end cap fitted.

Specialist techniques and processes

REVISED

Marking out

- Pencil or pens can be used to mark out on paper, card and foam board.
- Styrofoam, Corriflute and PVC foam can be marked with a thin permanent marker or chinagraph pencil.

Wastage/addition

- **Wasting** is the process of shaping material by cutting away unwanted parts to leave the desired shape.
- Addition is the adding of material by joining in some way, such as gluing pieces of thin card together to increase thickness.

> **Wasting**: the process of shaping material by cutting away waste material.

Cutting

- Scissors and craft knives are used for cutting paper and card.
- Foam board, PVC foam and Corriflute are too thick to be cut with scissors so must be cut using a craft knife.
- Styrofoam thicker than 10 mm can be cut using a serrated edged knife, bandsaw, hacksaw blade or hot wire cutter.
- Final shaping can be done by sanding and smoothing with files and abrasive paper.
- Laser cutters can be used to cut any 2D shape out of card, PVC foam, foam board or Corriflute.
- Die cutters are used to crease, perforate and cut card for mass production. The 'die' is a set of sharp metal blades shaped to the outline of the net and fixed to a backboard.
- The card or paper being cut is placed on a flat surface and the die is pressed onto the material which cuts it to the required shape.

Deforming/reforming

- **Deforming** is the process of changing the shape of a material by applying force, heat or moisture depending on the material.

Folding

- Paper and thin card can be folded easily by hand.
- Scoring will help fold thicker card and ensure a clean, sharp crease.
- Foam board can be folded by cutting through the foam in one of two ways:
  - Hinge cutting is where the foam board is cut part-way through so that the bottom layer of card acts as a hinge and the card can be folded backwards.
  - Vee cutting is where a v-shaped cut is made in the foam board and the material removed. This allows the foamboard to be folded inwards and gives a clean, tidy fold.
- PVC foam cannot be folded unless it is cut partway through, in a similar way to foam board.
- Corriflute can be folded by cutting a section of material away from the top layer between the flutes.
- Styrofoam cannot be folded.

> **Deforming**: changing the shape of a material by applying force, heat or moisture.

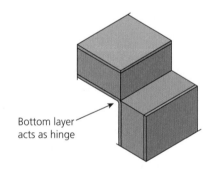

Bottom layer acts as hinge

Figure 7.6 Hinge cutting

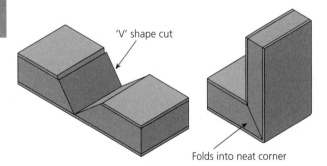

'V' shape cut

Folds into neat corner

Figure 7.7 Vee cutting

Joining

- There are a range of different adhesives available (see Table 7.5).
- Staples come in different sizes and strengths, from light duty ones for paper through to heavy duty staples which can pierce thick card, wood and MDF.
- Plastic rivets, also known as clic rivets, can be used to join thicker boards, including Corriflute and foam board together.

Table 7.5 Adhesives

Adhesive	Medium	Properties
Glue sticks	Paper and thin card	Easy to apply, inexpensive and mess free, can unstick over time, pieces can break off making surface lumpy, quick setting
Spray glues	Paper and thin card	Give a light even coating, can be permanent or temporary fixing, quick setting, can be messy
PVA glue	Thicker card, corrugated cardboard, foam board, Styrofoam	Dries clear, inexpensive, can be watered down
Hot glue guns	Thick card, Corriflute	Quick setting, can burn material and user, cool melt versions available, coloured glue sticks available
Cyanoacrylate glue (superglue)	Corriflute, PVC foam	Quick setting, very strong bond, short shelf life (one year approx.)
Polystyrene cement	Corriflute, PVC foam	Dries quickly, clear finish
Contact adhesive	Corriflute, PVC foam	Must be applied to both surfaces, needs to partly set before joining, instant bond, does not require clamping
Epoxy resin	Corriflute, PVC foam	Needs to be mixed with hardener, very strong bond, gives off strong fumes

Surface treatment and finishes

REVISED ☐

Paper

- Coatings can be applied to paper to improve the **opacity**, lightness, surface smoothness, **lustre** and colour-absorption ability.
- Cast **coatings** are when the wet paper is coated with china clay, chalk, starch, latex and other chemicals then rolled against a polished, hot, metal drum, creating a smooth, reflective and shiny surface that produces sharper, brighter images when printed on.

> **Opacity**: lacking transparency or translucence.
>
> **Lustre**: a gentle shine or soft glow.
>
> **Coating**: an additional outer layer added to a product.

Super calendering

- Super calendering is when paper passes through a calender or super calender. This is a series of rollers with alternately hard and soft surfaces that press the paper to create a smoother and thinner paper with a very high lustre surface.
- Super-calendered paper is primarily used for glossy magazines and high-quality colour printing.

Table 7.6 Types of paper finish

Type of paper	Properties	Used for
Cast-coated paper	Provides the highest gloss surface of all coated papers and boards	Labels, covers, cartons and cards
Lightweight coated	A thin, coated paper, which can be as light as 40 g/m²	Magazines, brochures and catalogues
Silk- or silk-matt-finished papers	Smooth, matt surface. High readability and high image quality	Product booklets and brochures
Calendered or glossy paper	Glazed shiny surface – can be coated and/or uncoated	Colour printing
Machine-finished paper	Smooth on both sides. No additional coatings applied after leaving the paper-making machine	Booklets and brochures
Machine coated	Coating applied while it is still on the paper machine	All types of coloured print
Matt-finished paper	Slightly rough surface prevents light from being reflected. Can be both coated and uncoated	Art prints and other high-quality print work

Card and board finishes

Varnish

- Varnish coatings enhance the look and feel of graphic products.
- Spirit varnish is used to give a high-shine finish that feels like plastic.
- Spot UV varnish is a special varnish that is cured or hardened by **UV light** during the printing process. Spot varnish is only used on certain areas of the paper or card to make it shinier and stand out.

> **Ultraviolet (UV) light**: outside the human visible spectrum at its violet end.

Hot foil application

- Hot foil application is used to produce metallic finishes such as gold or silver.
- It is often used for lettering on invitations and business cards.

Embossing and debossing

- Embossing and debossing give paper and card a 3D image that can be seen and felt.
- Embossing creates a raised area on the paper or card that stands out slightly.
- Debossing has the opposite effect and creates a sunken or lowered area.

Laminating

Laminating is usually done to finished documents to:

- improve their strength and resistance to bending, creasing or ripping
- waterproof the document, allowing it to be wiped clean and prevent it smudging or going soggy
- improve the appearance, making the document shiny
- increase the lifespan of the printed document.

Laminating involves applying a film of clear plastic between 1.2 and 1.8 mm thick to either one or both sides of paper or thin card. There are three methods of laminating a document.

Pouch lamination

- Pouch lamination uses thin clear plastic pouches coated on the inside with a thin layer of heat-activated glue.
- The document is placed inside the film pouch then fed through a laminating machine.
- The machine heats the pouch, activating the glue which seals the pouch together as it is pressed through the rollers encasing the document inside it.
- Pouch laminators can only do one document at a time so are ideal for small-volume items.

Thermal or hot lamination

- Where large volumes of laminated documents are needed, commercial lamination methods are required. Thermal lamination uses rolls of thin, heat-sensitive plastic film.

Cold laminating

- Cold laminating is done when only one side of the paper or card is to be coated.
- Cold laminating does not use heat, making it ideal for documents such as photographs.

Die cutting

- When mass producing products using paper and cardboard, a die cutter is used to crease, perforate and cut the card to shape.
- The 'die' is a set of sharp metal blades shaped to the outline of the net and fixed to a backboard.
- The die is pressed onto the material which cuts it to the required shape.
- Due to the high costs of making a die it is only used when producing large quantities of items.

Now test yourself

TESTED ☐

1 Describe some of the reasons why food packaging cannot always be recycled [4]
2 State two uses of Styrofoam and explain the properties that make it suitable for each purpose. [4]
3 Explain what is meant by the term 'super calendering'. [2]
4 Explain why some solvent-based adhesives are unsuitable for Styrofoam and foam board. [2]
5 Explain why paper can only be recycled a limited number of times and the effect recycling has on the quality of the paper. [3]

8 Natural and manufactured timber

Sources, origins, physical and working properties

Primary sources

- **Softwoods** mainly come from cool northern parts of Europe, Canada and Russia.
- **Hardwoods** are grown in Central Europe, West Africa, Central and Southern America.
- When trees reach maturity they can be **felled** and converted into planks.
- Newly-felled timber contains a lot of moisture and is known as **green timber**.

Seasoning

- Once **converted** into planks, timber must be seasoned to reduce its moisture content.
- In air seasoning, air flows around the stacked timber and gradually dries out the moisture. This process can take a number of years.
- In kiln **seasoning**, timber is housed in a kiln where steam is allowed to flow around the timber. The moisture in the steam is reduced which dries the timber. This is a much quicker process.

> **Softwoods**: timber that comes from coniferous trees and is generally less expensive than hardwoods.
>
> **Hardwoods**: timber that comes from deciduous trees and is generally harder than softwood.
>
> **Felling**: the process of cutting down trees.
>
> **Green timber**: timber that has just been felled and contains a lot of moisture.
>
> **Conversion**: the process of cutting a log up into planks.
>
> **Seasoning**: the process of removing moisture from newly-converted planks.

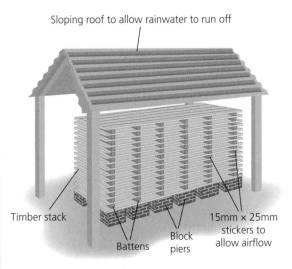

Figure 8.1 Air seasoning

Labels: Sloping roof to allow rainwater to run off; Timber stack; Battens; Block piers; 15mm × 25mm stickers to allow airflow

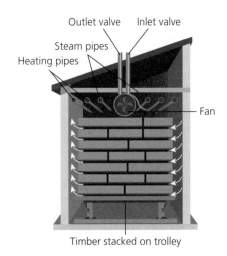

Figure 8.2 Kiln seasoning

Labels: Outlet valve; Inlet valve; Steam pipes; Heating pipes; Fan; Timber stacked on trolley

Defects in timber

- Shrinkage happens when the timber is allowed to dry out in an uncontrolled way. It can twist, warp, cup and/or bow.
- Splits occur in the ends of the timber if the drying out process is not controlled.
- Shakes are cracks in the timber that are a result of uncontrolled drying out of the timber. Thunder shakes can happen when a felled tree hits the ground or even during a thunderstorm!

- Knots form in timber where a branch has grown out of the tree trunk. Knots can cause several problems: they can weaken the timber; they can lead to decay and they can release resin which makes applying a finish to the wood very difficult.
- Fungal attack occurs when timber is left unseasoned. Fungus will cause disease and ruin the wood.

Types of natural timber and their working properties

- Natural timber is categorised into two groups – hardwoods and softwoods.
- Hardwoods come from deciduous trees that have broad leaves that generally fall in autumn.
- Hardwoods are generally harder, more expensive, more durable and take longer to grow than softwoods.

Table 8.1 Hardwoods

Hardwood	Properties	Common uses
Jelutong	A close-grained timber with a pale colour Medium hardness and toughness Easily worked	Pattern making
Beech	A hard, strong, close-grained timber with a light brown colour with distinctive flecks of brown Prone to warping and splitting Can be difficult to work	Furniture, children's toys, workshop tool handles and bench tops
Mahogany	A strong and durable timber with a deep reddish colour Available in wide planks Fairly easy to work but can have interlocking grain	Good-quality furniture, panelling and veneers
Oak	A hard, tough, durable, open-grained timber Can be finished to a high standard	Timber-framed buildings, high-quality furniture, flooring
Balsa	A very lightweight, soft and easily worked timber Pale in colour but weak and not very durable	Model making, floats and rafts

- Softwoods come from coniferous trees that have needle-type leaves, which they keep all year round.
- Softwoods are generally easier to work and, as they grow faster than hardwoods, are considered to be more sustainable than hardwoods.

Table 8.2 Softwoods

Softwood	Properties	Common uses
Western red cedar	Very resistant to weathering and decay Has a light reddish-brown colour with a close, straight grain Easily worked	Fencing, fence posts and cladding
Scots Pine	A straight-grained, light yellow-coloured timber Soft and easy to work Can be quite knotty	Interior joinery and furniture, window frames
Parana pine	Has a very distinctive open, straight grain Contains few knots and is strong and durable	Internal joinery and staircases

Types of manufactured timber and their working properties

Manufactured boards are commercially produced sheets of timber that offer advantages over natural timber:

- They are available in much larger sheets than solid timber (2440 mm × 1220 mm).
- They have consistent properties throughout the board.
- They are more stable than natural timbers, meaning they are less likely to warp, shrink or twist.
- They can make use of lower-grade timber, so can have environmental and economic benefits.
- They can be faced with a **veneer** or a laminate to improve their aesthetic appearance.
- Due to their consistent quality, they are well suited to CNC machining and volume production.

Manufactured boards fall into two categories:

- Laminated boards are produced by gluing large sheets or veneers together.
- Compressed boards are manufactured by gluing particles, chips or flakes together under pressure.

> **Manufactured boards**: sheets of timber that have been manufactured to give certain properties.
>
> **Veneers**: thin sheets of natural timber.

> **Exam tip**
>
> Learn the names and the properties of several natural timbers and manufactured boards. Make sure that you can suggest possible uses for them and be ready to identify them from a photograph.

Table 8.3 Manufactured timber

Manufactured board	Description	Properties
Medium-density fibreboard (MDF)	Made from compressed fine wood fibres bonded together with resin	This board is relatively inexpensive and has a flat, smooth surface
Plywood	Made from wood veneers glued together with alternating grain	Very strong, with a flat, smooth surface
Chipboard	Made from wood chips bonded together with resin	Inexpensive construction material. Limited strength
Hardboard	Made from compressed fine wood fibres bonded together with resin. Has one smooth side and one textured side	Very inexpensive material used for drawer bases and backs of wardrobes

Ecological and social footprint

- When correctly managed, timber is an ecologically-sound material with a good social footprint.
- Trees are a renewable natural material.
- Young trees convert more carbon dioxide into oxygen than older trees do, therefore felling old trees and replacing them with saplings (young trees) has environmental benefits.
- The conversion of timber into planks uses a limited amount of fossil fuels to power machinery when compared to the processing of materials such as metals and polymers.
- Timber is **biodegradable** and therefore has little effect on the environment if it goes to landfill.
- Timber products can be reused and are relatively easy to repair.

> **Biodegradable**: a material that will decompose into the earth.

Organisations such as the Forest Stewardship Council® (FSC) ensure that forests are correctly managed:

- Trees are replanted.
- Employment is made available to local workers.
- Workers are paid a fair wage and are given the correct safety clothing and training.
- Endangered wildlife is protected.

Deforestation

- If the felling of timber is not managed then **deforestation** occurs.
- Deforestation can lead to fertile soil being washed away, leading to a barren landscape.
- Deforestation causes the loss of habitats for wildlife.

> **Deforestation**: the removal of trees from an area of land.

Life cycle of timber products

Designers and manufacturers who use timber should be aware of the life cycle of the products they make – see Figure 8.3.

Typical mistake

When asked to describe the life cycle of a product made from natural timber, candidates will often start with the finished product and concentrate on how it is disposed of. You must start from its primary source, the tree.

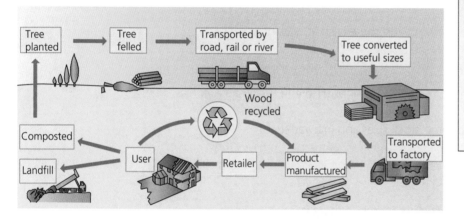

Figure 8.3 Life cycle of a wooden product

Selection of materials and components

REVISED

Functionality

- A garden bench needs to be strong, durable and weatherproof.
- A child's toy needs to be tough and free from splinters.
- Low-cost furniture needs to be flat, stable and easy to assemble at home.
- Different timbers have different properties:
  - O Pine is relatively inexpensive with an attractive grain and is available in long straight lengths making it ideal for larger prototypes.
  - O Beech is tough, durable and does not splinter, making it a suitable material for use in children's toys.
  - O Mahogany is strong with a deep red colouring that can be polished to a high gloss finish and is often used to manufacture smaller items such as jewellery boxes.

Aesthetics

- There is an infinite range of wood colours from sycamore (very pale) to ebony (black).
- Timber can be stained, painted and have a gloss, satin or matt finish applied to it.

- Timber is very tactile; it has a natural grain that can be sanded smooth or left quite rough.
- Timber is 'warm' to the touch.

Environmental factors

- Timber is environmentally friendly as it can be renewed by growing more trees.
- Timber can be recycled.
- It is biodegradable and will not harm the environment when it is disposed of.
- It is relatively easy to repair and most timbers have a good lifespan.
- Manufactured boards are less environmentally friendly as they have undergone additional manufacturing processes, and many contain adhesives that make recycling difficult.

Availability

- Softwoods are relatively quick growing and there is an abundance of renewable timber.
- Hardwoods take longer to grow.
- Most timber comes in stock sizes which makes designing and making easier to plan.
- Manufactured boards come in large stock sizes and in a variety of thicknesses.

True cost

- Softwoods are a relatively low cost natural timber; there is an abundant supply, it grows quickly and is easily converted into usable planks.
- Many manufactured timbers are also low cost; they can be made from low grade or recycled timber and are produced in high quantities.
- Hardwoods tend to be more expensive since they are slower growing and require a more involved process to convert them into a usable material.
- Some exotic hardwoods, such as burr walnut, can be very expensive. These are specialist timbers that are difficult to source and require expert attention to turn them into high quality usable timber.
- Labour-intensive manufacturing costs of bespoke hand-made furniture can increase the price of wooden products.
- Manufactured boards are suitable for mass production by CNC machinery that significantly reduces the cost of making.
- Manufactured timber is available in large sheets (2440 mm x 1220 mm) and in a variety of thicknesses, making it ideal for use in cost effective, self-assembly furniture.
- Bulk buying of timber can reduce the cost of the material.

Social factors

- Timber is an accessible, affordable material.
- It is used in all countries as a building and construction material, providing inclusive housing and furniture for everyone.

Cultural factors

- Different cultures have different needs and tastes regarding timber-based products.
- The Japanese make extensive use of bamboo as a timber-based construction material.
- In Northern Europe, log cabins are made from spruce.

Ethical factors

- Timber creates few ethical issues as it is a natural, renewable material.
- If timber forests are not managed, this can lead to deforestation.
- Deforestation can cause the loss of habitats for wildlife and can be a contributor to global warming.

Biodiversity

- Forests are a very biodiverse environment.
- Forests provide a habitat for many types of plants and wildlife.
- Animals, birds, insects, grasses and flowers live within forests and rely on them for their existence.

The impact of forces and stresses

REVISED

Timber can be used to give structure and strength to many products.

- A wooden chair leg must be strong enough to take the weight of the chair and the person who will sit on it. It must have good compressive strength.
- Oak is a particularly strong timber that can be used for the manufacture of beams in the construction of buildings.
- Beech is another strong durable timber that is used to make the benches that you use in the workshop.
- Scots pine is used to construct roof trusses. The timber is relatively lightweight and has good compressive and tensional strength, enabling it to take the weight of the roof.
- The strength of timber is also influenced by its cross-section. A thicker piece of timber is stronger than a thin cross-section of timber.
- Defects will negatively affect the strength of timber; knots will weaken timber and must be positioned so they do not compromise the structure.

Reinforcing and stiffening

- **Laminated timber** beams are an example of an engineered wood product. They are manufactured from layers of parallel timber laminations, normally of softwoods like pine or spruce, although hardwoods can be used.
- The sawn timbers are selected for strength before being glued together with the grain running in line with the laminates.
- **Manufactured boards** such as plywood are much stiffer than natural timbers. Plywood is a laminated board made up of several veneers of

wood glued on top of each other. Each layer is laid at a 90° angle to the last, so that the grain alternates in direction.

- The number of layers is always an odd number and the two outside surfaces always have the grain running in the same direction. This gives plywood a consistent strength across the whole board. By adding more layers, the strength of the plywood (now called multiply) is significantly increased.

Joining and fixing

- The method of fixing timber products together will affect the strength of the product.
- Nailing is a quick and economical method of fixing two pieces of wood together, but on its own offers little strength.
- Screwing timbers together offers greater strength and has the advantage of being a non-permanent fixing that can be taken apart if necessary.
- A butt joint is the simplest of joining methods, but is weak in comparison to a dovetail joint, which has the advantage of interlocking the wood.
- Adding a glue, such as PVA, to a wood joint significantly improves its strength and gives the reassurance of a permanent fixing.

Figure 8.4 Laminated beams above a Crossrail station in London

Stock forms, types and sizes

REVISED

Natural timber

- Natural timber that has come straight from the sawmill is known as rough sawn. It is generally used for construction work where appearance is not important.
- Timber that has had just its sides planed is known as planed both sides (**PBS**).
- Timber that has had all sides planed is known as planed square edge (**PSE**) or planed all round (**PAR**). This can be used for a wide variety of applications, including interior joinery.

PBS: planed both sides.

PSE: planed square edge.

PAR: planed all round.

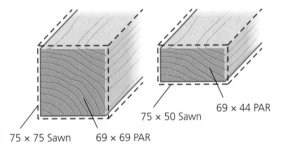

75 × 75 Sawn   69 × 69 PAR

75 × 50 Sawn

69 × 44 PAR

Figure 8.5 Typical timber planed sizes

Manufactured boards

- Manufactured boards such as medium-density fibreboard (MDF), plywood, hardboard and chipboard come in large sheets: 2400 mm x 1200 mm.
- The thickness of manufactured boards can range from 1 mm modelling plywood to 40 mm MDF, suitable for the core of a kitchen worktop.

Mouldings and dowelling

- Timber can also be supplied in a wide range of decorative mouldings. These are very useful when producing products such as picture frames. Dowelling is timber that is cylindrical and is useful for making wooden rails.

Veneers

- Veneers are thin sheets of wood that can be for constructional or decorative use.
- Construction veneers are typically used in the manufacture of plywood.
- Decorative veneers can be applied to manufactured boards to enhance their appearance.

Calculating the cost involved in designing timber-based products

- In addition to the cost of buying the timber you should also consider the additional cost of purchasing fixtures and fittings such as hinges, screws and handles.
- Different types of finish can also affect the final cost of a product. A simple wax finish is relatively cheap and quick to apply, whereas a high-quality finish such as French polish involves a very intricate process, requiring a high degree of skill and is much more time consuming, making it significantly more expensive.

Manufacturing to different scales of production

REVISED

One-off production

- Used to manufacture bespoke timber products.
- High-quality exotic timbers can be used.
- Products are manufactured by highly skilled woodworkers.
- Products are generally very expensive.
- The production method is very labour intensive and time consuming.

Batch production

- A range of identical wooden products can be made to a high consistency.
- Materials can be purchased in bulk, therefore reducing the cost.
- Machinery can be set up to manufacture in quantity, therefore saving time.
- Less skilled labour is required.
- The unique appeal of a 'one-off' product is lost.

> **One-off production**: only one, unique product is produced.
>
> **Batch production**: a number of identical products are produced.

Jigs and devices

A **jig** is a device that is specially made to perform a specific part of the manufacturing process.

Jigs are extremely useful when the process has to be carried out multiple times. They can be used when cutting, drilling, sawing and gluing. They have a number of very important **advantages**:

- they speed up the manufacturing process
- they reduce the risk of human error
- they reduce the unit cost of a part
- they make the process safer to carry out
- they increase the accuracy of the process
- they increase the consistency of the process.
- they reduce wastage.

There are some **disadvantages** of using jigs:

- they are only cost effective when large numbers of similar parts are required
- they increase the initial cost of the part
- they require a high level of skill to produce.

> **Jigs**: mechanical aids used to manufacture products more efficiently.

Mass production

- Timber products are manufactured in large quantities.
- Bulk buying of materials significantly reduces cost.
- Specialist machinery is used to increase consistency, accuracy and speed of production.
- An unskilled/non specialist workforce can be used.

Continuous flow production

- Timber products are made continuously for 24 hours a day, seven days a week.
- Highly specialised equipment and extensive use of computer-aided machinery (CAM) is used to manufacture the timber products.
- The process can be fully automated, deskilling the workforce who become involved in servicing and maintenance, rather than making.
- This requires a large initial investment and is only suitable where there is a high demand for the timber products.

> **Mass production**: large quantities of identical products are produced.
>
> **Continuous flow production**: identical products are being constantly made due to the high demand.

Issues with high-volume production

- Workers become deskilled and there is less employment as the machines take over manufacture.
- Timber products become very similar and lose their uniqueness.
- More energy is need to power factories, creating greater pollution for our planet.

Specialist techniques and processes

REVISED

Timber can be worked in a number of ways to produce a quality prototype.

Wastage and addition, deforming and reforming

- Wastage involves cutting, sawing and shaping timber to form a desired shape.
- Addition involves joining timber components together using wood joints and/or gluing.
- Deforming and reforming means producing shapes by processes such as laminating and steam bending.

Marking out

- Before marking out you should ensure that you have one face and one edge planed smooth. These are known as the 'face side' and the 'face edge'.
- A pencil and a ruler are used to mark out wood. A marking knife will give a more accurate cut line.
- A try square will give you an accurate 90° line to an edge. It will also allow you to check to see if an angle is at 90°.
- A mitre square produces an accurate 45° angle and a sliding bevel can be set to any angle.
- A marking gauge will produce a line parallel to an edge, while a mortise gauge will produce a double parallel line.
- A template can be used as a method of marking out irregular shapes on wood. You can also draw around it to use it as an aid to quantity production.

Sawing wood

- A tenon saw will cut straight lines in wood. Make sure the wood is firmly held in a vice or with a G-clamp.
- A hand saw is used for cutting large pieces of wood and a coping saw will cut curves.
- The coping saw can be difficult to use accurately and therefore you should always cut slightly away from your line.
- Bandsaws and scroll saws are mechanical saws that will speed up the process and can improve accuracy.

Shaping wood

- Wood can be shaped using a surform or wood rasp.
- A disc sander, belt sander and linisher are machines that use coarse glass paper to shape and smooth wood.
- Planes can be used to smooth and shape wood.
- Chisels are used to shape wood but are also very useful when cutting joints. Care should be taken when chiselling; always have your hands behind the cutting edge and securely clamp your work.

Drilling wood

- Drills will produce a circular hole in timber.
- Drilling machines can be a permanent feature in a workshop or handheld.
- Drill bits come in a wide variety of sizes.
- A hole saw can produce large holes; a Forstner bit can produce flat-bottom holes of varying sizes.

Joining wood

- There is a wide range of methods of joining wood.
- Carcase joints are used to make box-type constructions. Joints such as the 'butt joint' are quite simple to produce but are relatively weak. The 'dovetail joint' is difficult to produce but provides very good strength.

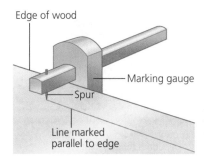

Edge of wood
Marking gauge
Spur
Line marked parallel to edge

Figure 8.6 A marking gauge

Figure 8.7 A mortise gauge

Figure 8.8 Sawing with a hand saw

Figure 8.9 Sawing with a tenon saw

Figure 8.10 Using a chisel to cut a housing joint

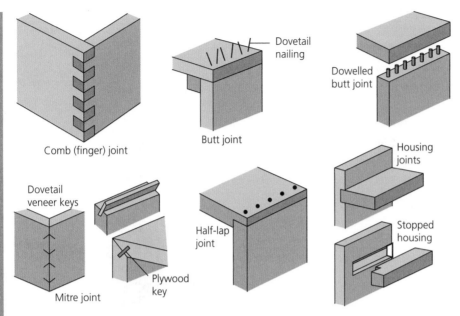

Figure 8.11 Carcase or box joints

- Stools tables and chairs can be constructed using stool joints.
- Wooden windows and doors are produced using frame joints.

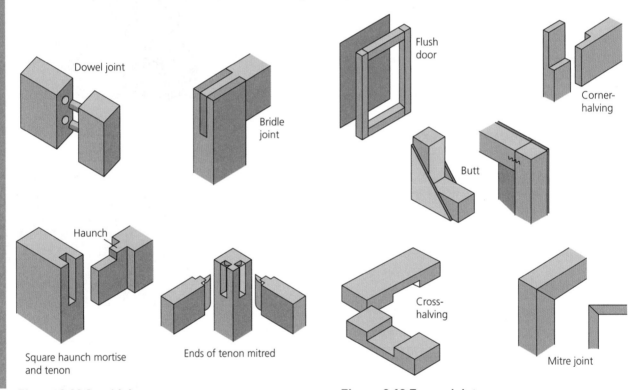

Figure 8.12 Stool joints

Figure 8.13 Frame joints

Adhesives

- The two most popular glues used when joining wood are PVA (polyvinyl acetate) and contact adhesive. PVA is a very strong glue but it takes quite a long time to dry. Contact adhesive is only a medium-strength glue but provides a quick joint.
- Epoxy resins can be used to join wood to other materials such as metals and polymers.

Woodscrews

- Woodscrews are a quick and convenient method of fastening two or more components together.
- Modern woodscrews are made from steel but have a protective coating to stop them from rusting.
- They are designed to be used with power tools such as a cordless drill for fast fixing.

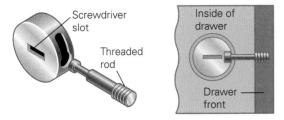

Straight slot   Phillips   Pozidriv

Figure 8.14 Common screwdriver slots

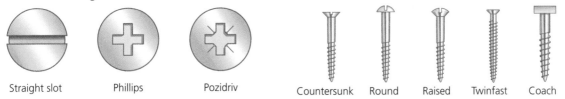

Countersunk   Round   Raised   Twinfast   Coach

Figure 8.15 Common screw heads

KD fittings

- Knock-down (KD) fittings are often used with 'flat-pack furniture'.
- They enable furniture to be sold unassembled, taken home in a flat cardboard box and assembled using simple tools.
- The concept of flat-pack furniture has greatly reduced the cost of buying furniture.

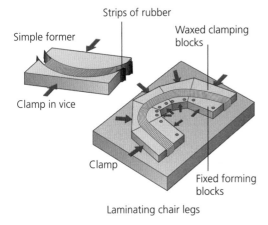

Figure 8.16 A cam lock fitting, used in flat-pack furniture

Lamination and steam bending

- Complex curves can be formed in wood by the processes of laminating and steam bending.
- Laminating involves gluing veneers of natural timber in between two halves of a mould.
- When steam bending timber, a solid section of natural timber is first steamed for several hours to make it pliable. It is then clamped into a mould and held for several hours until cool.

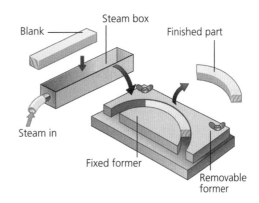

Figure 8.17 The process of laminating

Figure 8.18 The process of steam bending

Veneering

- Veneers are thin sheets of natural wood that can be applied to manufactured boards to enhance their appearance.

Computer-aided manufacture (CAM)

- A laser cutter can cut thin sections of natural timber and manufactured boards from a computer-aided design (CAD) drawing.
- A laser cutter can also be used to etch an image or working onto the surface of timber.
- A 3D router can be used to produce a 3D solid timber product from a computer-aided design (CAD) drawing.
- Close grained hardwoods, mdf and modelling foams work best with the 3D router.

Surface treatment and finishes

REVISED

- Surface finishes are generally applied to timber-based materials to protect the material, enhance its appearance and to increase its durability.
- A finish is required on most natural timbers as they need to be protected from the weather.

Surface preparation

- Preparing the surface is an essential part of the finishing process.
- The timber should be sanded using a variety of grades of sandpaper to remove any marks and smooth the surface.
- The surface should be free from dust, dirt and oil.

Types of finish

- A wood stain will change the colour of the timber but offers little protection.
- A wood preservative soaks into the timber and protects it from both moisture and insect attack. Preservatives can also include a stain. They are generally used on garden sheds and fencing.
- Tanalising is a commercial method of applying a preservative. The timber is pressure treated with the preservative. Tanalised timber is used extensively in the manufacture of patio decking.
- Varnishes offer good protection to timber. It is normally used as a clear coating but is also available in a range of colours.
- Oils, such as Danish oil and Teak oil, are easy to apply to timber and give a reasonable level of protection. They are not particularly long lasting and will require recoating each year.
- Wax polishes give a low level of protection and are generally used on top of a varnished surface. French polishing is a specialised method of applying polish and is used only for high-quality furnishings.
- Paints give a high level of protection to timber and are available in a wide range of colours. Most paints require an extra level of preparation. If there are knots in the timber they must first be primed, the surface of the timber must then have an undercoat before the final gloss/satin/matt coat is applied.
- A veneer is often applied to manufactured boards to improve their appearance.

Now test yourself

1 Name a suitable natural timber for use as a wooden spatula and explain your choice.   [3]
2 Explain why newly-felled timber must be seasoned.   [4]
3 Explain the advantages of using KD fittings to both the customer and the manufacturer of wooden furniture.   [6]
4 Use notes and sketches to describe how to produce a bend in wood using the steam bending technique.   [6]
5 Give two reasons why it is necessary to apply a finish to patio decking.   [2]

9 Ferrous and non-ferrous metals

Sources, origins, physical and working properties

Primary sources

Most metals are found in the Earth's crust and are embedded in rock known as **ore**.

- Ore can be opencast mined, underground mined or even dredged from rivers.
- Iron comes from an ore called **haematite**, which can be found in countries such as Brazil, Australia and South Africa.
- Ore must be **smelted** to release the metal from the rock.
- Iron is extracted from haematite by heating the ore to a very high temperature in a blast furnace.
- Aluminium is smelted from its ore (**bauxite**) in a reduction cell.

> **Ore**: rock which contains metal.
>
> **Haematite**: ore containing iron.
>
> **Smelting**: the process of extracting metal from ore.
>
> **Bauxite**: ore containing aluminium.

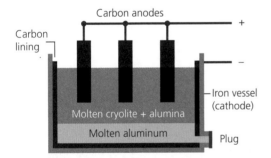

Figure 9.1 The reduction cell

Types of metal and their working properties

Ferrous metals

All ferrous metals contain iron and when alloyed (mixed) with carbon they produce steel.

Table 9.1 Common ferrous metals

Ferrous metal	Composition	Properties	Common uses
Cast iron	Iron and 3.5% carbon	Hard surface but has a brittle soft core, strong compressive strength, poor resistance to corrosion, 1200°C melting point, good electrical and thermal conductivity, cheap	Vices, car brake discs, cylinder blocks, manhole covers
Mild steel	Iron and 0.15%–0.35% carbon	Good tensile strength, tough, malleable, poor resistance to corrosion, 1500°C melting point, good electrical and thermal conductivity, cheap	Car bodies, nuts, bolts, and screws, RSJs and girders
Medium-carbon steel	Iron and 0.35%–0.7% carbon	Good tensile strength, tougher and harder than mild steel, poor resistance to corrosion, 1500°C melting point, good electrical and thermal conductivity	Gardening tools and springs
High-carbon steel	Iron and 0.70–1.4% carbon	Hard but also brittle, less tough, malleable or ductile than medium-carbon steel, poor resistance to corrosion, 1500°C melting point, good electrical and thermal conductivity	Screwdrivers, chisels, taps and dies

Heat treatment of ferrous metals

- The properties of ferrous metals can be altered by the use of heat.
- **Annealing** involves heating the metal to red heat and then allowing it to cool very slowly. This makes the metal as soft as possible.
- To **harden** ferrous metal, heat it to red heat and then cool it as quickly as possible in cold water.
- **Tempering** involves heating the steel to a known temperature and then allowing it to cool naturally. This removes the brittleness of hardened steel.
- The process of case hardening can harden the surface of a ferrous metal. The metal is heated to red heat and then placed in a high-carbon compound where it soaks up some of the carbon. This carbon-rich coating can then be hardened by heat treatment.

> **Annealing**: a method of heat-treating metal that makes it as soft as possible.
>
> **Hardening**: a method of heat-treating metal that makes it hard but brittle.
>
> **Tempering**: a method of heat-treating metal that reduces brittleness.

Non-ferrous metals

Non-ferrous metals do not contain iron.

Table 9.2 Common non-ferrous metals

Non-ferrous metal	Composition	Properties	Common uses
Aluminium	Pure metal	Lightweight, soft, ductile and malleable, a good conductor of heat and electricity, corrosion resistant, 660°C melting point	Aircraft bodies, high-end car chassis, cans, cooking pans, bike frames
Copper	Pure metal	Extremely ductile and malleable, an excellent conductor of heat and electricity, easily soldered and corrosion resistant, 1084°C melting point	Plumbing fittings, hot water tanks, electrical wire
Silver	Pure metal	A soft, precious metal that is extremely resistant to corrosion, an excellent conductor of heat and electricity, 961°C melting point, expensive	Often used as jewellery

Heat treatment of non-ferrous metals

The properties of non-ferrous metals can also be altered by the use of heat. The main difference is that hardening and annealing non-ferrous metals happens at a much lower temperature.

Alloys

An alloy is a metal that is produced by combining two or more elements together to produce a new metal with enhanced properties.

Table 9.3 Common alloys

Alloy	Composition	Properties	Common uses
Stainless steel – ferrous alloy	Alloy of steel also including chromium (18%), nickel (8%) and magnesium (8%)	Hard and tough, excellent resistance to corrosion, 1510°C melting point, good electrical and thermal conductivity	Sinks, cutlery, surgical equipment, homewares
High-speed steel – ferrous alloy	A medium-carbon alloy that also contains tungsten, chromium and vanadium	Very hard, resistant to friction, can only be sharpened by grinding, 1540°C melting point, good electrical and thermal conductivity	Lathe cutting tools, drills, milling cutters
High-tensile steel – ferrous alloy	A low-carbon steel that also contains chromium and molybdenum	A steel with a very high yield strength, 1540°C melting point	Used to reinforce concrete
Brass – non-ferrous alloy	Alloy of copper (65%) and zinc (35%)	Strong and ductile, casts well, corrosion resistant, 930°C melting point, conductor of heat and electricity	Castings, forgings, taps, wood screws
Bronze – non-ferrous alloy	Copper 80–90%, tin, aluminium, phosphorous and/or nickel in varying amounts	Reddish-yellow in colour, harder than brass, corrosion resistant, 1200°C–1600°C melting point	Castings, bearings and gears
Pewter – non-ferrous alloy	Tin 85–90%, antinomy, copper	Malleable with a low melting point (170°C–230°C), easy to work	Castings, beaten metalwork
Duralumin – non-ferrous alloy	Alloy of aluminium (90%), copper (4%), magnesium (1%), manganese (0.5–1%)	Strong, soft and malleable, excellent corrosion resistance, lightweight, 660°C melting point, conductor of heat and electricity	Aircraft structure and fixings, suspension applications, fuel tanks

Ecological and social footprint

Metal has both positive and negative impacts on the ecology and society.

- The mining and processing of metal ores creates employment for many people in countries all over the world.
- The metal-based manufacturing industry can be a very hazardous environment; employers should ensure that workers are provided with safety training and equipment.
- The mining of metal ore can create scars on the landscape and lead to deforestation, both of which can have a negative effect on the habitat of wildlife.
- Metal products are generally non-biodegradable and some can even pollute the ground.
- Metal products can be repaired and recycled.

Life cycle of metal products

Recycling metal can save as much as 90 per cent of the total energy used to manufacture metal products from the original source.

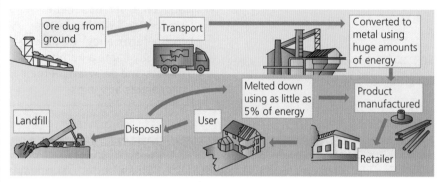

Figure 9.2 Life cycle of metal products

Selection of materials and components

Function and aesthetics

- Metals are selected for both their functional and aesthetic properties.
- They are more difficult to work than most materials due to their inherent hardness but can be manufactured to greater accuracy.
- Table 9.4 outlines the functional and aesthetic qualities of metals.

Table 9.4 Functional and aesthetic properties of ferrous and non-ferrous metals

Metal	Aesthetics	Functionality
Aluminium (duralumin)	Easily cast into unique shapes, can be polished to a mirror like finish or be coloured with a vivid finish known as anodising	Excellent strength-to-weight ratio, easy to cut, weld and joined by various methods, good resistance to corrosion
Copper	Easy to shape by beating, has a reddish-brown finish that can be highly polished, has the unique feature of going green when left outdoors and unprotected	Easily worked, good conductor of heat and electricity, malleable, ductile and easily joined by soldering
Brass	Easily cast into unique shapes, has a yellowy-brown colour that can be highly polished	A harder, more durable material than copper, a good conductor of heat and electricity
Bronze	Often used by sculptors to cast into works of art, has a deep reddish-brown colour that can be highly polished	A hard, durable material that is very resistant to corrosion
Pewter	Very easily cast into shape due to its low melting point	Soft metal with a relatively low strength
Silver	Can be polished to a high-quality finish, widely used in the manufacture of elegant jewellery	A hard, durable metal that can be easily joined by soldering, excellent resistance to corrosion, expensive
Cast iron	Nearly always cast into a variety of shapes, rusts if unprotected	A hard, durable, strong, heavy metal
Mild steel	Can easily be worked into shape, rusts if left unprotected, can receive a wide variety of finishes/platings/coatings	Tough, durable, strong and malleable, relatively easy to work
Medium-carbon steel	Harder to work into shape than mild steel, will rust if left unprotected but can accept a variety of finishes	Very tough, durable and strong

Environmental factors

- Metals come from ore which is a non-renewable resource.
- The processing of ore into pure metal uses a lot of energy. Most of this energy comes from non-renewable fossil fuels.
- Some metals will pollute the ground if they go to landfill.
- Most metals are relatively easy to repair, have a good lifespan and can be recycled.

Availability

- Metal is readily available in a variety of stock forms (see Stock forms, types and sizes below).
- Metal comes in stock sizes of length, width, thickness (gauge), diameter and weight.

True cost

- The cost of metals can vary quite considerably. Common metals such as steel are relatively inexpensive. Semi-precious metals such as copper, tin and lead are more expensive, while precious metals such as gold and silver are very expensive.
- Metals are suitable for batch, mass and continuous methods of manufacture, which reduces the unit cost of everyday products such as aluminium drinks cans.
- Bespoke items of jewellery, made by hand using precious metals, are very expensive.

Social and cultural factors

- Constructional metals such as steel are very accessible and affordable and therefore they are extensively used by all.
- Precious metals such as gold and silver are only affordable by the wealthy members of society.

Ethical factors

- Metals are a finite resource and must be recycled at the end of their life or they will not be available for future generations.
- The ELVD (End of Life Vehicles Directive) is a European directive aimed at ensuring that vehicles are correctly dealt with at the end of their usable life.

The impact of forces and stresses

REVISED

- Most metals can withstand an impressive amount of force and stress.
- Steel is an excellent construction material as it has great strength. It has exceptional tensile strength and this allows steel cables to be used to hold up suspension bridges.
- Softer metals such as lead are malleable and can withstand being beaten into shape without cracking. For example, look at where a chimney enters the roof of a house. The silver metal around the chimney is made from lead and has been beaten into shape.
- Copper is a ductile metal and can withstand being drawn out into fine wire. The metal inside most of the electrical cables in your house is copper.

Improving the properties of metals by hardening and tempering

- High-carbon steel can be made even harder by a heat treatment process known as hardening.
- The steel is heated with a brazing torch, in a brazing hearth, until it is red hot. It is then quickly cooled in water, which hardens the steel.
- The hardening process also makes the steel brittle. To remove the brittleness the steel must once again be heated, but this time to a specific temperature between 230 and 300°C.

Colour	Temp. °C	Hardness	Typical uses
Light straw	230	Hardest	Lathe tools, scrapers
Dark straw	245		Drills, taps and dies, punches
Orange/brown	260		Hammer heads, plane irons
Light purple	270		Scissors, knives
Dark purple	280		Saws, chisels, axes
Blue	300	Toughest	Springs, spinners, vice jaws

Figure 9.3 Tempering colours and temperatures

Stock forms, types and sizes

REVISED

- It is important to consider the common stock forms and sizes of metal when designing and making metal components. This will save you time and effort as metal is a hard material to cut and shape.
- Some of the most popular stock forms of metal are shown in Figure 5.3 in Topic 5.

Manufacturing to different scales of production

REVISED

One-off production

- Typically used to manufacture bespoke metal products such as jewellery from precious metals.
- Products can be manufactured using highly skilled metalworkers.
- Products are usually very expensive.
- The production method is very labour intensive and time consuming.

Batch production

- Products such as mountain bikes can be manufactured in large numbers.
- Materials can be purchased in bulk therefore reducing the cost.
- Machinery can be set up to manufacture in quantity, saving time.
- Less skilled labour is required.

The use of jigs

- Jigs are extensively used in the industrial manufacture of metal products.
- Drilling jigs allow a series of holes to be accurately drilled in exactly the same place, time after time.
- Welding jigs allow metal components to be held in place ready for welding.
- For more information on jigs see Topic 8.

Mass production

- This allows metal products such as car body panels to be produced in large quantities.
- Bulk buying of metals significantly reduces cost.
- Specialist machinery and the increased use of computer-aided manufacture (CAM) are essential features of high volume production. They increase consistency, accuracy and the speed of production.
- An unskilled/non-specialist workforce can be used.

Continuous flow production

- Metal products that are in very high demand such as food cans are made continuously for 24 hours a day, seven days a week.
- Highly specialised equipment and extensive use of CAM is used to manufacture the metal products.
- The process can be fully automated, deskilling the workforce who become involved in servicing and maintenance.
- This requires a large initial investment and is only suitable where there is a high demand for metal products.

Issues with high-volume production

- Workers become deskilled and there is less employment as the machines take over manufacture.
- Metal products become similar and lose their uniqueness.
- More energy is need to power factories, creating greater pollution.

Specialist techniques and processes

REVISED

Wastage/addition

There are specific marking out, cutting, shaping and joining methods used with metal-based materials and it is important to know which to use.

Marking out

- A scriber acts as a pencil when marking out on metal. On bright metal it is useful to use marking-out fluid to make it easier to see the lines.
- An engineer's square produces a 90° line to an edge and a centre punch will mark the centre of a hole before drilling.
- Outside and inside callipers measure the outside and inside of a bar and tubing.
- Digital micrometres and Vernier callipers can measure to 1/100th of a millimetre.

Sawing

- The hacksaw is the most common saw used for cutting straight lines. A junior hacksaw is a small version used for cutting smaller metal parts.
- Bandsaws, jigsaws and even a coping saw can all be fitted with metal cutting blades.

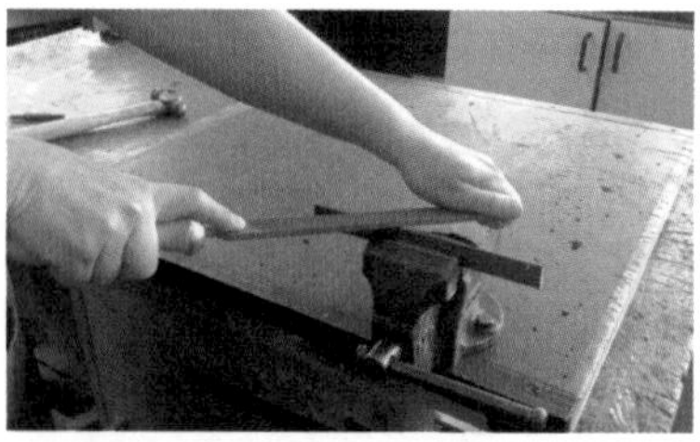

Shaping

- Metal can be shaped using a range of different types of file. Files are graded depending on their roughness; these are known as rough cut, second cut and smooth cut.
- Cross filing is an effective method of shaping metal. You should use the full length of the file and remember that the file only cuts on the forward stock.
- Draw filing smooths the edges of metal. Place the file across the edge of the metal while holding the blade rather than the handle. Then move the file forwards and backwards to achieve a flat, smooth edge.

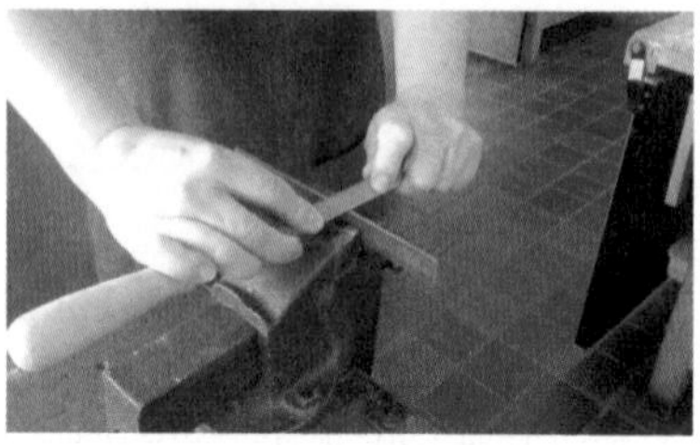

Drilling metal

- Metal must be centre punched before drilling to ensure that the drill bit does not slip.
- The speed of the drilling machine should be set to the size of the drill and the hardness of the metal. Soft metals and small drills require a fast speed while hard metal and large drills need a slower speed.
- Drilling metal is normally done using a pillar drill, with the metal firmly held in a machine vice.
- Drill bits will produce a regular circular hole in metal.
- A countersunk bit will produce a countersunk hole.
- A **pilot hole** is a small hole produced to guide a large hole.
- A **tapping hole** is produced before an internal screwthread is formed.
- A **clearance hole** is a hole that is produced to allow a bolt to slip through.

Deforming/reforming

Non-permanent methods of joining metal

- Nuts, bolts and washers are the most common method of non-permanently fixing metal components together.
- Machine screws allow metal components to be non-permanently fixed together.

Figure 9.4 Cross filing

Figure 9.5 Draw filing

Figure 9.6 A countersink drill bit

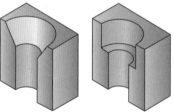

Countersunk hole   Counterbored hole

Figure 9.7 A countersunk hole and a counterbored hole

Figure 9.8 Nut, bolt and washer

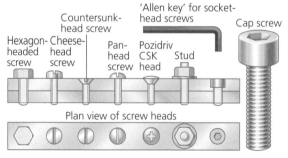

Figure 9.9 Types of machine screw

Permanent methods of joining metal

- **Riveting** is a mechanical method of permanently fixing metal parts together. A hole is drilled through the metal components, then a rivet is placed in the hole and hammered into shape. When pop riveting, a rivet gun is used to form the rivet.

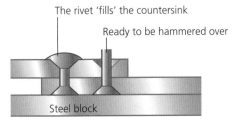
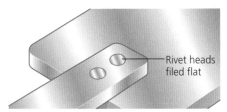

The rivet 'fills' the countersink

Ready to be hammered over

Rivet heads filed flat

Steel block

Figure 9.10 The riveting process

- **Soft soldering** can be used when attaching copper pipe fittings together. The joint is first cleaned, a paste flux is applied, and then it is heated with a torch. The solder then melts and flows into the joint. Solder used for plumbing is made from tin and copper.

- **Hard soldering** is similar to soft soldering but is used for joining precious metals together such as gold and silver.

- **Brazing** is a similar process to hard soldering but uses brass as a solder to attach steel components together.

Figure 9.11 Soldering copper piping

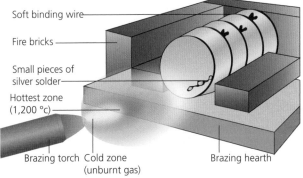

Soft binding wire

Fire bricks

Small pieces of silver solder

Hottest zone (1,200 °c)

Brazing torch   Cold zone (unburnt gas)   Brazing hearth

Figure 9.12 The hard soldering process

- **Welding** uses heat to melt the surface of a metal. Once the metal is in a molten state it pools together and any gaps are filled with a filler rod.

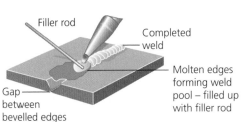

Filler rod

Completed weld

Molten edges forming weld pool – filled up with filler rod

Gap between bevelled edges

Figure 9.14 Oxyacetylene gas welding

Clean steel surface

Flux to help keep the join clean and help the metal to 'flow'

Brazing rod has melted and 'flowed into' joint

Figure 9.13 Brazing

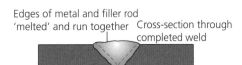

Edges of metal and filler rod 'melted' and run together   Cross-section through completed weld

- **Epoxy resins** can be used to glue metals together to form a low to medium strength joint. Both surfaces of the metal should have a clean, keyed surface. (Keying involves lightly scratching the surface with an abrasive paper and gives the adhesive something to bond to.) Epoxy resin consists of two parts, an adhesive and a hardener; these should be mixed together in equal quantities, applied to the metal surface and then clamped until the glue sets.

Machining

Centre lathe

- The centre lathe allows cylindrical metal bars to be machined. The ends of the bar can be 'faced off' making them flat and smooth after being cut with a hacksaw.
- The diameter of the bar can be reduced by a process known as 'parallel turning' or a cone can be formed by 'taper turning'.
- The ends of metal can also be accurately drilled using a drill attached to the tailstock of a centre lathe.

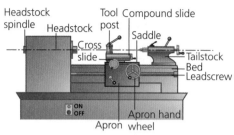

Figure 9.15 Centre lathe

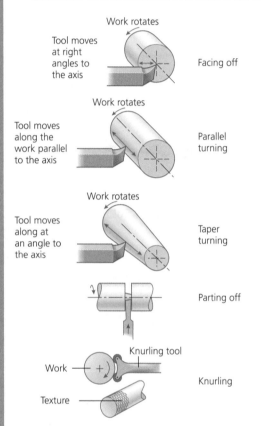

Figure 9.16 Lathe operations

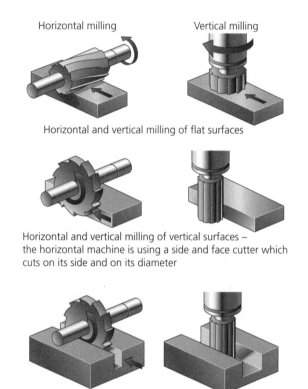

Horizontal and vertical milling of flat surfaces

Horizontal and vertical milling of vertical surfaces – the horizontal machine is using a side and face cutter which cuts on its side and on its diameter

Horizontal and vertical milling of slots

Figure 9.17 Milling operations

Milling machines

- **Milling** machines will machine flat surfaces and allow slots and grooves to be cut into metal.

CNC lathes and milling machines

- Metal is ideally suited to using CNC lathes and milling machines.
- A CNC lathe can parallel turn, face off, taper turn and even cut screwthreads.
- A CNC milling machine can cut slots, grooves, machine edges and create flat smooth surfaces in metal.
- Both machines follow CAD drawings to a high level of accuracy and are faster and more consistent than machining by traditional methods.

Plasma cutter

- These perform similar operations to a laser cutter but can cut and engrave quite large thicknesses of metal.

> **Milling**: cutting grooves and slots into metal.
>
> **Turning**: a method of producing cylinders and cones using a centre lathe.

Figure 9.18 A plasma cutter

Surface treatment and finishes

- Surface treatments and finishes are applied to metals to enhance their appearance and to offer them protection.
- Ferrous metals require a finish to be applied otherwise they will rust.
- Non-ferrous metals may be given a finish to enhance their appearance but precious metals such as gold and silver are usually just polished.
- Finishing always begins with surface preparation. The surfaces should be sanded smooth and be free from dirt, dust and oil.

Dip coating

- Dip coating covers the metal in a layer of polyethylene.
- The metal is fully cleaned and then placed into an oven and heated to a temperature of 200°C.
- The metal is then dipped into a fluidising bath for a few seconds.
- The polyethylene sticks and melts onto the hot surface leaving a smooth shiny coating.

Powder coating

- This is an industrial equivalent of dip coating.
- The metal is fully cleaned and then placed into an oven and heated to a temperature of 200°C.
- The metal is then sprayed with a layer of powdered polyethylene from an electrostatic spray gun.
- The electrostatic spray gun ensures an even coating over the metal.
- The metal is then placed back into the oven to cure.

Galvanising

- This is an industrial process that involves coating the metal in zinc.
- The metal is fully cleaned and then dipped into a bath of molten zinc.
- The zinc provides a durable and corrosion resistant barrier.

Anodising

- Anodising is an electrolytic finishing process that is specific to aluminium.
- The aluminium is fully cleaned by a series of mechanical and chemical methods.
- The aluminium is then placed into a chemical bath where a current is passed through the metal, forming an oxidised layer.
- The oxidised layer hardens the surface and provides a corrosion resistant protective layer.
- Coloured dyes can be added to the chemical bath which colour the oxidised layer.

> **Typical mistake**
>
> If a question asks you for a detailed description of how to prepare and apply a finish to a metal component, make sure that you include details of any surface preparation that may be required. Also include details of any health and safety issues that may arise.

Figure 9.19 Galvanised steel safety barrier beside a road

Enamelling

- Enamelling can be used to decorate jewellery or to protect household appliances.
- Enamelling involves coating metal with powdered glass that is then fused to the metal by heating it in a kiln to a temperature above 800°C.

Figure 9.20 Enamel jewellery

Oil blacking

- Oil blacking produces a very thin layer of black oxide on the surface of ferrous metals.
- Oil blacking is used on machine tools and components to prevent them from rusting.
- A ferrous metal component is heated and then soaked in an oil bath.

Painting

- Painting metals prevents corrosion and enhances their appearance.
- Surface preparation is very important, and metals should be free from dust, dirt and oil.
- A primer coat is applied followed by several layers of coloured gloss paint.
- Paint can be applied by brush or be sprayed on to the metal surface.

> **Typical mistake**
>
> Marks can be lost when candidates provide superficial, vague responses to questions that ask for a detailed response.

> **Now test yourself** TESTED ☐
>
> 1 Use notes and sketches to describe the life cycle of a steel can. [6]
> 2 Name two methods of permanently joining metal together. [2]
> 3 Name two non-permanent methods of joining metal together. [2]
> 4 Explain the purpose of a pilot hole. [2]
> 5 Use notes and sketches to describe how you would dip coat a metal component. [6]

10 Thermosetting and thermoforming polymers

Sources, origins, physical and working properties

REVISED ☐

Synthetic polymers

- Most polymers are made from crude oil and are known as **synthetic polymers**.
- Crude oil is found throughout the world, with the biggest deposits being in the Middle East and in Central and South America.
- It is extracted from the ground by drilling and pumping it to the surface. It is then transported to an oil refinery to be processed.
- There are three stages to the processing of crude oil:
  1 **Fractional distillation** – crude oil is boiled to produce gas, which is vented off where it condenses to form products such as gas, petrol and naphtha, which is used to produce polymers.

> **Synthetic polymers:** polymers that are sourced from crude oil.
>
> **Fractional distillation:** the processing of crude oil into naptha.

2 **Cracking** – naphtha is heated and condensed again to break it down into ethylene, propylene and butylene, which are needed to produce polymers.

3 **Polymerisation** – monomers are linked together to form polymers. Ethylene is linked to form polyethylene (PE) and propylene is linked together to form polypropylene (PP). Butylene can be mixed with other polymers to form products such as holt melt glue.

> **Cracking**: the processing of naptha into monomers.
>
> **Polymerisation**: the blending of different monomers to create a specific polymer.
>
> **Natural polymers**: polymers that are sourced from plants.

A simple monomer

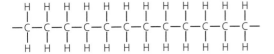

The structure of the polymer polyethylene

Figure 10.1 Monomers and polymers

Natural polymers (biopolymers)

- Biopolymers are made from plant-based materials such as sugar beet and corn starch.
- Being plant-based means that biopolymers are renewable

Thermoforming polymers

Thermoforming polymers can be softened by heating, then formed and shaped by a wide variety of processes.

> **Thermoforming polymer**: a polymer that can be reheated and reformed.

Table 10.1 Properties and uses of thermoforming polymers

Thermoforming plastic	Properties	Common uses
Acrylic (PMMA)	Hard, excellent optical quality, good resistance to weathering, scratches easily and can be brittle, excellent thermal and electrical insulator, good plasticity when heated	Car-light units, bathtubs, shop signage and displays
High-density polythene (HDPE)	Hard and stiff, excellent chemical resistance, excellent thermal and electrical insulator, good tensile strength, good plasticity when heated	Washing up bowls, buckets, milk crates, bottles and pipes
Low-density polythene (LDPE)	Flexible and tough, waterproof and suitable for all moulding techniques, excellent thermal and electrical insulator, good plasticity when heated	Plastic bags and refuse sacks
Polyvinyl chloride (PVC)	Hard and tough, good chemical and weather resistance, low cost, can be rigid or flexible, excellent thermal and electrical insulator, good plasticity when heated, good tensile strength	Pipes, guttering, window frames
Polypropylene (PP)	Tough, good heat and chemical resistance, lightweight, fatigue resistant, excellent thermal and electrical insulator, good plasticity when heated, good tensile strength	Toys, DVD and Blu-ray cases, food packaging film, bottle caps and medical equipment

Thermoforming plastic	Properties	Common uses
Polycarbonate (PC)	Tough, durable and impact resistant, good resistance to scratching, excellent thermal and electrical insulator, good plasticity when heated	Safety glasses, safety helmets
Extruded polystyrene foam (XPS) Styrofoam	Lightweight, easy to work, good thermal insulation	Thermal insulation barriers in the construction industry, modelling material
Expanded polystyrene (EPS)	Lightweight, easy to mould, good impact resistance, excellent thermal insulator	Packaging, disposable cups and plates
Nylon	Hard, tough and resistant to wear, a low coefficient of friction, excellent thermal and electrical insulator	Bearings, gears, curtain rail fittings and clothing

Thermosetting polymers

Thermosetting polymers differ from thermoforming polymers in that once they are set, they cannot be reformed by heating.

> **Thermosetting polymer**: a polymer that cannot be reformed with heat.

Table 10.2 Properties and uses of thermosetting polymers

Thermoset	Properties	Common uses
Epoxy resin	Excellent thermal and electrical insulator, good chemical and wear resistance, can be brittle	Adhesives such as Araldite®, PCB component encapsulation
Melamine formaldehyde (MF)	Stiff, hard and strong, excellent resistance to heat, scratching and staining, excellent thermal and electrical insulator	Kitchen work-surface laminates, tableware
Urea formaldehyde (UF)	Stiff and hard, excellent thermal and electrical insulator	Electrical fittings, toilet seats, adhesive used in MDF

Additives

There are a number of additives than can be blended with polymers to enhance certain properties.

- Polymerisers are added to polymers to enhance their flexibility.
- Pigments change the colour of a polymer.
- Fillers are added to polymers to increase their bulk and reduce their cost.
- Flame retardants can be added to polymers to prevent or slow down the rate at which they burn.

> **Exam tip**
>
> Be able to name several polymers, give their properties and suggest possible uses.

Ecological and social footprint

- Polymers come from crude oil which is a non-renewable fossil fuel.
- The processing of crude oil into a polymer uses a lot of energy and creates pollution.
- The environment and wildlife can be seriously affected by spillages of crude oil.
- Most polymers are non-**biodegradable** and if they go to landfill they will last for many, many years.
- Polymers that find their way into the oceans are being ingested by fish and are making their way back into the food chain.
- Polymers are not easy to repair but they are durable and can have a good lifespan.

> **Biodegradable**: a material that will degrade/decompose by natural means.

> **Exam tip**
>
> Make sure that you know the difference between a thermoforming polymer, thermosetting polymer and a natural polymer (biopolymer).

Now test yourself answers at www.hoddereducation.co.uk/myrevisionnotes

- Many polymers can be recycled.
- Biopolymers are sourced from plants and are therefore renewable.

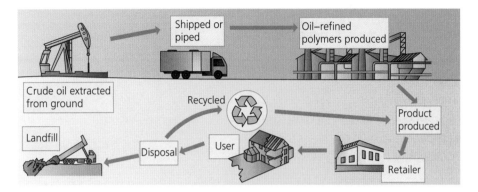

Figure 10.2 Life cycle of a polymer product

Selection of materials and components

Functionality

- Thermoforming polymers can be moulded using industrial machinery, making them ideal for high volume production.
- Polymers do not corrode, therefore they do not need to be protected from water. They are not affected by many chemicals.
- They are a good thermal and electrical insulator.
- Most polymers are very tough and durable.
- Polymers are self-coloured, waterproof, chemical resistant and often have a perfect shiny surface. This reduces many of the finishing processes associated with other materials.
- Thermosetting polymers are unaffected by heat once they have been moulded.

Aesthetics

- Thermoforming polymers can be moulded with the use of heat to produce a wide variety of shapes.
- Polymers have a natural high-gloss finish and can be transparent, translucent or opaque.
- The appearance of a polymer can be easily changed by adding a coloured pigment.

Environmental factors

- Sourcing, processing and the manufacture of polymer-based products has many negative effects on the environment.
- If polymers go to landfill they can stay around for many, many years. If they find their way into the oceans they can enter the food chain.
- Many polymers are durable; they can easily be cleaned and reused.
- Synthetic polymers come from a non-renewable source and are not considered to be environmentally friendly. However, many are recyclable and durable which limits their negative effect on the planet.

True cost

- While polymers are more expensive than many timber or metal-based materials, the true cost of a polymer-based product can be significantly cheaper as they can be easily manufactured in high volumes using industrial moulding machines.

Social factors

- The introduction of modern/smart polymers and advanced methods of manufacture means that new products are now available to everyone.
- There is an increasing pressure for people to own the latest model of product. This can lead to debt or a feeling of failure if you cannot afford it.

Cultural factors

- The needs, beliefs and tastes of different cultures must be taken into account when designing polymer-based products. Examples include:
  - In the late 1940s, developments allowed affordable polymer storage containers to be produced by a company known as 'Tupperware'. Throughout America and then Europe, Tupperware products were in high demand and people even held Tupperware parties!
  - Today, companies will sell the same smartphone in different colours to suit the particular preferences of different countries.

Ethical factors

- The use of synthetic polymers can have a negative effect on our environment. We must decide whether a 'greener'/renewable material would be a better choice.
- Polymers should be reused, repaired or recycled at the end of the initial life cycle.
- Many polymer-based products are manufactured in large industrial factories where making a profit is a major factor in how products are made. However, it is essential that the workers should be paid a fair wage and work in a safe environment.

The impact of forces and stresses

REVISED

- Polymers are naturally waterproof, tough, durable and are excellent electrical and thermal insulators.
- The cross-sectional area and the depth of a polymer **extrusion** has a significant effect on the strength of a polymer product. A deep extrusion with a large cross section will produce a strong polymer product.
- Polymers can be moulded with reinforcing webs and support material to increase the strength of the product.
- Acrylic (PMMA) is a brittle material with poor bending strength and therefore should not be used it where it has to withstand a load.
- ABS is very similar to acrylic but has good strength and impact resistance. Car indicator lenses and children's toys are made from ABS.
- Glass reinforced plastic (GRP) has excellent strength and can be moulded to form the body of boat hulls.
- Carbon fibre reinforced plastic (CFRP) has excellent strength, is relatively light in weight and is used to produce the body panels of F1 racing cars.

> **Extrusion:** a length of polymer with a consistent cross section.

Figure 10.3 Carbon reinforced polymer (CFRP) artificial running blade

Stock forms, types and sizes

- Polymers are available in a wide range of forms, colours and thicknesses.
- A big advantage of using polymers is that they can be bought self-coloured with an immaculate high gloss finish.

Table 10.3 Stock forms

Stock form	Description
Sheet	The most commonly used sheet polymer in the school workshop is acrylic (PMMA). The sheet comes in regular sizes measuring $1200 \times 600 \times 3$ mm
	Other polymer sheet materials include High Impact Polystyrene (HIPS). This is available in a size that fits most vacuum-forming machines. $475 \times 274 \times 1$ mm
Extrusions	Typical extrusions include rod and tube. Specific extrusions include products such as curtain rails
Granules/pellets	Mainly used in industrial processes such as injection moulding and extrusion
Powders	Used in the fluidising bath as part of the plastic dipping process
	In industry they would be used in powder coating and also used with thermosetting polymers in compression moulding
Foams (closed cell)	Available in large sheets (1200×600 mm) and extensively used in school for model making
Foams (open cell)	Usually obtained in a can and sprayed into a cavity. Has good thermal, buoyancy and sound-insulating properties
Films	A very thin sheet that is usually sold in rolls
Filament	Polylactic acid (PLA) is sold in rolls and is used in many rapid prototyping machines
Liquid	A number of polymers are available in a liquid form. Epoxy resins are sold in liquid form and only become solid when mixed with a chemical catalyst

Figure 10.4 Polymer extrusions

Figure 10.5 Polymer granules

Calculating the cost of polymer-based products

- The cost of a polymer-based product is influenced by the type of polymer and the method of production.
- The cost of a product made from acrylic (PMMA) that is to be laser cut will be calculated from the surface area of PMMA needed plus the power and time consumed by the laser cutter.

- The cost of a prototype produced in polylactide acid (PLA) by a 3D printer will be calculated by the length of filament needed plus the power and time consumed by the 3D printer.
- The cost of a product produced on an industrial injection moulding machine from polypropylene (PP) will be calculated from the volume of PP needed plus the power and time consumed by the injection moulding machine.

Manufacturing to different scales of production

REVISED

One-off production

- An early example of a product that is produced is known as a 'one-off'. Its function is to be able to visualise and test the design.
- A one-off can also be a bespoke product made for a particular purpose. An example could be a Formula 1 racing car made from carbon reinforced polymer (CFRP).
- One-off products are generally expensive, time consuming to produce and require a highly skilled workforce.

Batch production

- This is when a limited number of identical products are produced.
- Machinery is often used in the manufacturing process.
- Batch production allows materials to be bought in bulk and a less skilled workforce to be used. This helps to reduce the unit cost of batch-produced products.

The use of jigs

A jig is a device that is specially made to perform a specific part of the manufacturing process. Jigs are extremely useful when the process must be carried out multiple times. They can be used when cutting, drilling, sawing and gluing.

Advantages of using jigs:

- They speed up the manufacturing process and reduce the risk of human error.
- They increase the accuracy and consistency of the process.
- They reduce wastage and the unit cost of a product

Disadvantages of using jigs:

- They are only cost effective when large numbers of similar parts are required.
- They increase the initial cost of a product.
- They require a high level of skill to produce.

High-volume production

- This involves the constant manufacture of products 24 hours a day, 7 days a week.
- This would normally be carried out using fully automated machinery.
- Used for products for which there is very high demand, such as water bottles.

Computer-aided manufacture (CAM)

- Computer-aided machinery (CAM) is used when polymer-based products need to be manufactured in high volumes.
- This requires a large initial investment in new factories and state-of-the-art machines.
- It is only cost effective when there is a very high demand for identical polymer-based products.

Processes in manufacturing polymer products

There are a number of processes that are used to manufacture polymer-based products in high volumes. These are:

- **blow moulding**
- injection moulding
- **vacuum forming**
- **press forming**
- compression moulding.

All these processes use the same principles of manufacturing.

- The polymer is heated and becomes soft and pliable.
- The polymer is then blown, sucked, drawn or pressed into a die or mould.
- The polymer takes the form of the die or mould.
- The polymer is then cooled.
- The product is removed from the die or mould and then trimmed and finished.

> **Blow moulding**: a method of shaping a thermoforming polymer by heating and blowing into a dome.
>
> **Vacuum forming**: a method of shaping a thermoforming polymer sheet by heating and sucking around a former.
>
> **Press forming**: a method of shaping a thermoforming polymer by heating and pressing into a mould.

Specialist techniques and processes

REVISED

Polymer-based materials utilise a range of tools and equipment that are also common to both wood and metal.

Wastage/addition

Marking out

- A spirit-based pen will draw onto a polymer surface without scratching it.
- Paper templates allow a design to be transferred to the surface of the polymer and will give extra protection to the shiny surface of the polymer.
- Keeping the protective sheet on a polymer for as long as possible will help keep the surface clean and free from scratches.
- A laser cutter can be used to etch marking-out lines onto the surface of most polymers.

Holding polymers

- When working with polymer-based materials, they can be held in a vice or with a G-clamp. Care should be taken to protect the surface.

Sawing

- Most metal and woodworking saws will cut polymers. The coping saw will allow you to cut around curves in acrylic (PMMA).

> **Typical mistake**
>
> When asked to describe a manufacturing process such as vacuum forming, candidates will often produce weak, inaccurate diagrams. Make sure that you can draw accurate, fully labelled diagrams of the main forming processes of vacuum forming and blow moulding.

- A scroll saw is a mechanised coping saw and will speed up the process.
- A craft knife will cut thin polymer sheets such as the high impact polystyrene (HIPS) sheets used for vacuum forming.

Shaping and forming polymers

- Polymers can be filed and sanded into shape by traditional methods using metalworking files and abrasive papers such as 'wet or dry' paper.
- Thermoforming polymers can be formed using a strip heater to produce a bend along a straight line.
- An oven can be used to soften a thermoforming polymer before it is shaped in a mould.
- Most cutting and shaping tools used in the manufacture of metal products can be used when making polymer-based components.
- A disc sander will speed up the process of shaping polymers.

Drilling polymers

- Polymers can be drilled using regular metalworking drill bits.
- Using a pillar drill will improve the accuracy of drilling but a cordless drill has the advantage of being able to drill in remote locations.
- A **pilot hole** is a small hole drilled before a larger hole is drilled. This will improve the accuracy of drilling and prevents the build-up of heat that would melt the polymer.
- A **clearance hole** is a hole that is drilled larger than the size of the threaded part of the bolt. This allows the threaded part of the bolt to slip through the component before fastening with a nut.
- A **tapping hole** is a hole that is drilled before an internal screwthread is cut.
- A **countersunk hole** is produced to accept a countersunk screw.
- A **counterbored hole** is a hole that has been enlarged to accept the head of a bolt.

Deforming/reforming

Temporary joints

- Polymers can be temporarily fastened together using nuts and bolts. The main difference in fixings designed specifically for use with polymers is that they generally have larger heads to displace the pressure over a larger area.
- Self-tapping screws and panel trim fixings provide a quick and efficient way of fixing polymer components together, for example, the way that plastic panels in a car are fixed in place.
- Polymer sheets can be riveted together using aluminium rivets, pop rivets or bifurcated rivets. Care must be taken not to apply excessive force as the polymer sheet can be easily damaged.
- Traditional butt and flush hinges can be used with polymers. However, some polymers such as expanded polystyrene (EPS) can be formed with an integral (built in) hinge.
- Traditional catches can be used with polymers, but integral catches can be built into the design.

Figure 10.6 A strip heater

Figure 10.7 Pop rivets and a rivet gun

Figure 10.8 Expanded polystyrene (EPS) food container with integral hinge

Figure 10.9 Food containers with handle and catch

Permanent joints

- Tensol cement (dichloromethane and methyl methacrylate) is a clear solvent adhesive that is specially formulated to glue polymers such as acrylic (PMMA).
- Certain polymers can be welded together using a hot air gun to melt the surfaces of the polymers. The seam on a plastic bag is welded together using a heated clamp.

Machining
The centre lathe and the milling machine

- A metal working centre lathe and a milling machine can perform all the same operations on polymer-based materials as they can on metal-based materials.
- See Topic 9 for more information on turning and milling operations.

Computer-aided manufacture (CAM)

- Polymers are ideally suited to volume production using CAM processes.
- A vinyl cutter uses a computer-controlled blade to cut self-adhesive vinyl by following a computer-aided design (CAD).
- A laser cutter uses a laser beam to cut and etch into certain polymers such as acrylic (PMMA) following a CAD drawing.
- A 3D router will mill slots and profiles into most polymers and is most often used with modelling foams such as Foamex.
- CNC lathes and CNC milling machines will perform the same operations as centre lathes and milling machines but are computer controlled following a CAD drawing.
- A 3D printer can produce a complete 3D polymer product from a 3D CAD drawing.

Figure 10.10 CNC lathe machining a component

Vacuum forming and blow moulding

Both these processes make use of a vacuum-forming machine to produce thin walled polymer shapes such as yoghurt cartons and clam shell packaging.

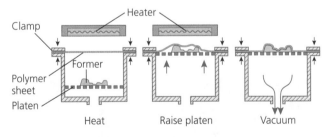

Figure 10.11 The vacuum-forming process

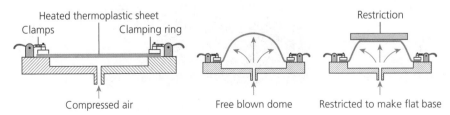

Figure 10.12 The blow-moulding process

Press forming

Thicker 3D shapes can be produced by press forming.

- An acrylic (PMMA) sheet is warmed in an oven until soft.
- The sheet is placed over the 'plug'.
- The 'yoke' is placed over the top of the acrylic and pressure is applied.
- The acrylic sheet will take the shape of the mould.
- The acrylic sheet is then left to cool, removed from the mould and trimmed.

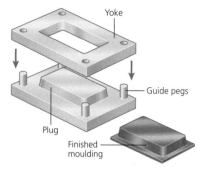

Figure 10.13 A press-forming mould

Surface treatment and finishes

REVISED

- Polymers are self-coloured and usually have an immaculate, high-quality finish.
- They are quite resistant to wear and tear and are not affected by water or by many chemicals.
- Therefore, they require little finishing.

Polishing

- The cut edges of a polymer can be brought back to a high-quality finish by draw filing, sanding with 'wet or dry' paper and then polishing using a metal polish or by buffing on a buffing wheel.

Printing

Decoration and texture can be applied to the surface of a polymer by the pad or screen-printing method.

Vinyl decals

Self-adhesive vinyl decals can be stuck onto the surface of a polymer-based product.

Figure 10.14 Vinyl decals

Textured finishes

A textured finish is generally created at the time of moulding. The finish is integrated into the shape of the mould. Textured finishes are often found on polymer-based products to give grip to the otherwise slippy, shiny surface.

Figure 10.15 Draw filing acrylic (PMMA)

Figure 10.16 Polishing acrylic (PMMA)

11 Natural, synthetic, blended and mixed fibres, and woven, non-woven and knitted textiles

Sources, origins, physical and working properties

REVISED

Choosing the most suitable fabric for a product is vital. Factors that need to be considered include:

- the source of the fibre and its properties
- how the fibre has been spun and made into yarn
- how the fabrics have been **constructed** from yarns or fibres
- applied finishes.

Woven, non-woven and knitted textiles

Woven fabric construction

- Weaving is done on a loom using warp and weft yarns.
- **Warp** yarns run along the length of fabric, called the **straight grain**.
- **Weft** yarns run horizontally across the fabric, called the **cross grain**.
- Weft yarns interlock with warp yarns in different formations creating variations in types of weave.
- The **selvedge** is the factory-finished edge of the fabric.

> **Fabric construction**: the way a fabric has been made.
>
> **Warp**: yarns that run along the length of fabric.
>
> **Straight grain**: indicates the strength of the fabric in line with warp yarns.
>
> **Weft**: yarns that run across the fabric.
>
> **Cross grain**: horizontally across the fabric from edge to edge.
>
> **Selvedge**: the sealed edge of the fabric.

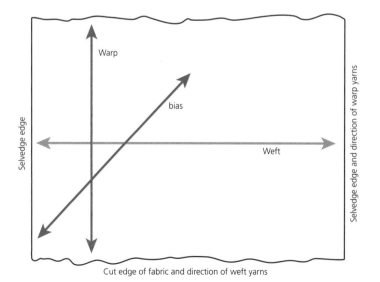

Figure 11.1 Warp, weft and selvedge

Table 11.1 Characteristics of woven fabrics

Woven fabric	Characteristics
Plain weave	The simplest structure and most commonly used weave Variations can be achieved through different thickness and textures of yarn, different colour combinations in the yarns and how closely the yarns are packed together
Twill weave	Recognised by the diagonal lines created by the weave Variations include herringbone and chevron Twill weave produces a strong, heavy and more durable fabric
Satin weave	Has a smooth, shiny, lustrous appearance created by the floating yarns in the weave A disadvantage is that it snags easily due to the structure of the floating yarns
Pile weave	Has a raised surface formed by tufts or loops in the weave which can be cut as in velvet, or left as in towelling Pile weave fabrics are hardwearing because of the thickness created by an extra loop in the yarn Pile fabrics such as corduroy have a directional surface so all pattern pieces have to be laid and cut in the same direction to avoid shading

Figure 11.2 A plain weave structure

Figure 11.3 A twill weave structure

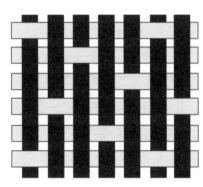
Figure 11.4 A satin weave structure

Non-woven fabrics

- **Bonded fabrics** are made directly from fibres, which makes them cheaper to use.
- Pressure and heat or adhesives are applied to a web of fibres to bond them together.
- Bonded fabrics have limited uses but are often used in disposable products such as surgical masks.
- **Felted fabrics** are non-woven fabrics made by applying pressure, moisture, heat and friction to staple fibres to bind them together.
- They are usually made from wool or acrylic fibres. They are easy to cut and will not fray but can stretch out of shape when wet. They are ideal for craft projects.

Figure 11.5 Hand-made felt is made by applying pressure, heat, moisture and friction

Knitted fabrics

- There are two types of knitted fabric: warp knitting and weft knitting.
- Knits are created by a series of interlocking loops in the yarn.
- Knitted fabrics are warmer than woven fabrics as they trap air within the loops.

Table 11.2 Knitted fabrics

Weft knit fabrics	Warp knit fabrics
Made from a single continuous yarn, constructed from horizontal rows of interlocking loops Can be hand-made or constructed on industrial machines Has an obvious right and wrong side Can snag If part of the yarn is damaged, cut or pulled they can unravel Can stretch easily Can lose shape	Made on automated machines from multiple yarns that interlock vertically Identical on both sides Do not run or unravel More flexible than weft knits Have some stretch Hold their shape well and can be cut to shape when making products

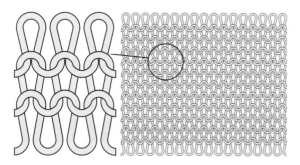

Figure 11.6 The weft knitting structure has a series of loops made from a single yarn that interlocks horizontally

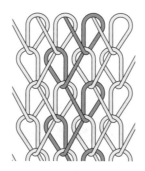

Figure 11.7 Warp knitting consists of multiple yarns interlocking vertically

Laminated and coated fabrics

- Fabrics are laminated to combine the performance characteristics of the fabrics used to make a superior fabric. Examples include neoprene as used in wetsuits.
- Laminated fabrics are either:
  - held together with an adhesive or
  - either a polymer film or a layer of foam is heated and pressed onto the fabric it is to be joined to.
- Gore-Tex® is a laminated fabric with a **hydrophilic membrane** as one of the inner layers.
- A coated fabric has a polymer, usually polyurethane or polyvinylchloride (PVC), applied to it which is then fixed in an oven. PVC-coated cotton and faux leather are two of the most well-known examples.

Figure 11.8 A wetsuit made from laminated neoprene

Fabric specification

A **fabric specification** sets out the fabric requirements for a product for a specific end purpose.

The physical and working properties of fibres, method of construction for the fabrics and applied finishes should be carefully considered before selecting an appropriate material to use.

Fibres

- Fibres are the raw materials of textiles and come from natural or manufactured sources.
- They are made from chemical units called polymers, formed from smaller single units called **monomers** which link together creating long chains.

Hydrophilic membrane: ability to repel and release moisture.

Fabric specification: sets out the requirements of the fabrics needed for a product.

Monomer: a molecule that can be bonded to others to form a polymer.

- Fibres are classed as:
  - long continuous **filaments**, for example, polyester, nylon and acrylic
  - short staple fibres, for example, cotton, linen and wool.
- The shape of fibres affects how the fibre feels or the **handle** (softness) and its lustre (shine).
- Fibre blends bring together fibres of different types to improve functionality, cost or appearance of the blended/mixed fibres.
- Common fibre blends include polyester cotton, cotton and elastane, wool and acrylic, silk and viscose.

> **Filament**: a very and fine slender thread.
>
> **Handle**: how a fabric feels when handled.
>
> **Crimp**: the waviness in a fibre.

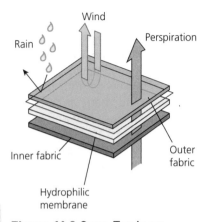

Figure 11.9 Gore-Tex is an example of a laminated fabric and is widely used for high performance clothing.

Table 11.3 Structure and description of fibre types

Fibre	Description
Cotton	Short staple fibres with a slight twist Smooth surface prevents air being trapped – poor insulator Inner cavity allows moisture to be absorbed
Linen	Short staple fibres Smooth surface prevents air being trapped – poor insulator Has a light shiny lustre which prevents soiling The structure has cavities making it highly absorbent
Silk	The only natural long continuous filament Consists of two long protein bundles closely packed together Highly absorbent making it very cool to wear
Wool	Fibres have a natural **crimp** (waviness) and are coated with scales The crimp allows air to be trapped so it is a good insulator The scales can hook together when wet causing shrinkage The natural grease in the fibres makes wool water repellent
Polyester	It can be engineered for different purposes – versatile A long flat filament that does not hold water – poor absorbency Can be crimped in the manufacturing process, allowing it to trap some air and improve insulation

Spinning

- Spinning is the process of twisting fibres together to make a yarn. The individual fibres are quite weak but gain additional properties when they are spun into yarns.
- There are two ways in which fibres are spun:
  - S twist (anticlockwise)
  - Z twist (clockwise).
- A tight twist squeezes out air, making the fibres closer together, resulting in a yarn that is strong but not warm. A loose twist allows more air to be trapped, making the yarn warmer but much weaker.

> **Exam tip**
>
> The central cavity within cotton and linen fibres enable these fibres to soak up and store moisture. When discussing the properties of cotton, it is important to note that the fibres are absorbent in the first instance. The ability to wick away moisture makes them appear breathable and cool to wear. Absorbency is the important property.

Fancy yarns

- Twisting multiple yarns together creates textured or novelty yarns with irregular surfaces and varying thicknesses, such as Bouclé.
- These add texture and surface interest when knitted or woven.

Quilting

- English quilting consists of three layers: a top layer, a plain lower layer and a layer of polyester wadding sandwiched between them.
- Reasons for quilting include:
  - insulation – air is trapped in between the layers which will keep the wearer warm
  - decoration – adding surface interests
  - functional reasons – reinforcement where added protection is needed.

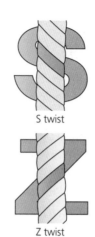

Figure 11.10 Spinning yarns – S twist, Z twist

Ecological and social footprint

Impact on the environment

- Crops such as cotton and linen are intensively farmed, which requires extensive areas of land and a large volume of water. Farmers use pesticides and fertilisers which can damage the ecosystem.
- Crops can be farmed organically using natural fertilisers and pesticides, but this still requires a substantial amount of water.
- Farming livestock for wool production can lead to land being damaged by over-grazing and insecticides used for protection from mites and ticks. Livestock produce the greenhouse gas methane which contributes to global warming.
- Non-renewable resources are needed for transporting raw materials and products and the extraction of synthetic fibres. This contributes to our carbon footprint.
- The textiles industry can cause water pollution through contaminated waste from processing and waste water from the washing, drying, dyeing and printing of fabrics. This affects fish in rivers and streams as well as drinking water supplies, which affects people's health.

Biodiversity

- Biodiversity is the balance of different creatures and species living side by side in a particular habitat, all being part of the same **ecosystem**.
- Damage or destruction of the ecosystem is caused by using land to grow textile fibre crops, taking water for crop production, infrastructure, pollution as a result of manufacturing and the disposal of textile waste.

Ecosystem: the natural and delicate balance of plant life, soil and animals that live interdependent lives.

Changing society's view of waste and recycling

- An estimated 350,000 tonnes of used clothing goes into landfill in the UK each year, yet most fibres and fabrics can be easily recycled.
- Fibres cannot withstand endless recycling and some will go to landfill at the end of their useful life.
- Synthetic fibres are non-biodegradable but are recyclable.

Social footprint

Working conditions

- The textile industry is known for being an exploitative industry with workers in developing countries often being subjected to long hours and low pay.
- Work is often carried out in dangerous conditions, while those who live in close proximity to the textile industry are also affected by the contamination of land and water supplies.
- **Fast fashion** has added to the exploitation of some workers.
- A **throwaway culture** and demand for lower retail prices leads to cuts in manufacturing costs which can mean lower pay for workers.

Life cycle

- Natural fibres are biodegradable and most will decompose naturally.
- However, they will release CO_2 during decomposition and fibres treated with dyes and chemicals will release these into the ground.
- Synthetic, polymer-based fibres decompose very slowly, taking up to 1000 years to break down. While doing so they release toxic and greenhouse gases into the atmosphere.

> **Fast fashion**: a recent trend copied from the catwalks, often low quality and low in price.
>
> **Throwaway culture**: refers to the culture of buying affordable fashionable items, wearing or using them for a short period of time before disregarding them or throwing them away; little thought is given to the environmental impact.

Selection of materials and components

REVISED

Mechanical and physical finishes

Table 11.4 Mechanical and physical finishes

Process	Aesthetic and functional effect
Brushing	Wire brushes are passed over the surface of the fabric to raise the fibres This produces a soft, fluffy surface, improving its appearance and enhancing insulation
Calendering	Fabric is passed through heated rollers to give it a smooth, more lustrous surface, enhancing its aesthetic qualities Moiré is a variation of this finish and gives a wavy watered effect
Glazing	A similar process to calendering but with stiffeners or resins added to the finish to make it more permanent
Embossing	Heat and pressure is applied to fabric as it passes through engraved rollers; this leaves an imprint or slightly raised surface pattern in the fabric

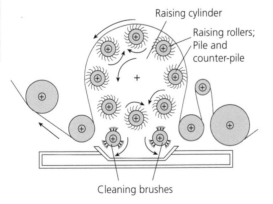

Raising cylinder

Raising rollers; Pile and counter-pile

Cleaning brushes

Figure 11.11 Wire brushes raise the fibres to produce a soft fluffy surface

Chemical finishes

Table 11.5 Chemical finishes

Process	Aesthetic and functional effect
Mercerising	Caustic soda is used to make the fibres in the fabric swell, leaving a more lustrous, stronger fabric This process only works for cellulose fibres
Crease resistance	A resin coating is applied to the fabric, stiffening the fibres and making products easier to care for Treated fabrics dry more quickly but reduce the ability to absorb moisture
Flame resistance	Chemicals such as Proban® are applied to the surface of fabrics as a liquid coating which reduces the fabrics' ability to ignite and burn
Bleaching	Bleaching removes any natural colour and is used to prepare fabric for dyeing and printing
Stain resistance	Teflon™ and Scotchguard™ are fabric protectors mostly used on clothing and home furnishings
Anti-static	A chemical-based product is applied to the fabric to prevent the build-up of electrostatic charge also known as static electricity
Water repellence	Silicon is applied as a semi-permanent finish which repels water Applying a fluorochemical resin makes fabric water repellent and wind resistant Teflon™ and Scotchguard™ are water-repellent finishes Coating with PVC, PVA or wax also repels moisture
Shrink resistance	A chlorine-based chemical smooths out the scales on the wool fibre which prevent them locking together during washing, preventing shrinkage
Moth proofing	This finish repels moths who feed off the keratin found in wool fibre

Biological finishes

Table 11.6 Biological finishes

Process	Aesthetic and functional effect
Stone washing	Stones added to industrial washing machines give a distressed 'worn out' look to fabrics. Often applied to denim jeans

Responsibilities of designers and manufacturers

Designers and manufacturers should consider the following:

- Most applied **finishes** use energy and water as well as chemicals and toxins which are hazardous to the environment and the health of the textile workers.

- Chemical finishes applied to fabrics reduce the fibre's ability to fully degrade; chemical traces damage the ecosystem and impact on biodiversity.

- New finishing chemicals are being developed that are reusable, do not require the use of water and are biodegradable.

> **Exam tip**
>
> To demonstrate knowledge and understanding, know how finishes are applied to fabrics and be able to explain the purpose of the finish, particularly in relation to the end product.

> **Typical mistake**
>
> Quite often answers relating to environmental issues are not fully explained. You will not gain marks by simply stating something is bad for or has a negative impact on the environment. Always try to give a specific example in your response and a full explanation of the impact.

> **Fabric finishes**: these are added to fabrics to improve their aesthetics, comfort or function. These finishes can be applied mechanically, chemically or biologically.

The impact of forces and stresses

Reinforcing and stiffening textiles

- Textile fabrics vary in strength depending on:
  - the fibre content: for example, polyester is a much stronger fibre than wool
  - the way a yarn has been spun: a tighter twist in spinning leads to a stronger yarn
  - the way the fabric has been constructed: twill weave, for example, is a much more stable weave than satin weave.
- Designers use a number of methods to reinforce and strengthen textile fabrics where more structure is needed in the final product.
- Geotextiles used in construction are extremely strong fabrics. They are often made from polypropylene or polyester fibres which allow the geotextile to perform as intended.
- Products such as loaded rucksacks or tents are subjected to additional force and stresses during use and need to be reinforced. Aluminium or carbon fibre frames are used for this purpose.
- Some seams are clearly much stronger than others, for example, a double stitched seam is stronger than a plain seam. Double stitched seams are used in workwear or sportswear for this reason.
- Quilting, laminating and bonding are also methods of reinforcing and strengthening fabrics.
- Layering textiles will add structural support, as well as insulating and making products more comfortable to wear.

> **Exam tip**
>
> Almost all textile products have some sections that are reinforced or strengthened to improve their structural integrity allowing them to function. The more products you analyse during your studies the better your understanding will be.
> Use this information when answering exam questions.

Table 11.7 Method of layering

Method of layering	Effect
Interfacing	A non-woven, bonded fabric that can be sewn or ironed on to the outer fabric or lining Provides support, stiffens and reinforces fabrics to improve the shape and add structure
Interlining	A layer of fabric added between the fabric and the lining of a garment Adds a layer of insulation and supports the structure of a garment
Lining	Usually lightweight fabrics such as polyester, satin or nylon Improves comfort Hides seams and construction methods Can be a design feature and may add warmth to the garment An underlining on sheer fabric might add body or provide opacity to prevent a garment being see-through

> **Interfacing**: a bonded fabric, which when fused to a different fabric adds strength.
>
> **Lining**: a layer of fabric in the same shape as the outer layer which adds support and conceals the construction processes.

> **Typical mistake**
>
> An interlining and a lining serve different purposes and should not be confused. A lining is visible on the inside of a garment, while the interlining is concealed between the lining and the outer fabric.

Now test yourself answers at www.hoddereducation.co.uk/myrevisionnotes

Method of layering	Effect
Boning	Plastic or metal boning strips are sewn into reinforced seams to support the fabric, preventing creasing and buckling Corsets are worn to accentuate and cinch the waist, giving a more desirable figure Medical corsets are used to treat back pain or protect people with spinal or internal injuries by restricting movement
Piping	Piping inserted into seams adds structure to the shape of a product, helps with durability and can be decorative Piping can be bought ready-made or it can be made by wrapping a piece of bias binding around a cord then inserting into a seam

Stock forms, types and sizes

REVISED

Stock forms

- Standard widths for textile fabrics include: 90 cm (interfacings or linings), 115 cm, 150 cm and 200 cm.
- Cotton sheeting at 240 cm wide is also available from some specialist fabric retailers.
- 'Fat quarters' are pre-cut fabric pieces measuring 45 cm x 55 cm, and are ideal for small projects.

Common fabric names

Table 11.8 Common fabric names

Fabric name	Example use
Denim	Jeans, dresses
Voile	Curtains
Chiffon	Evening wear, lingerie
Corduroy	Trousers, skirts, jackets
Velvet	Evening wear, jackets
Lace	Decorative, clothing

Fabric name	Example use
Cotton poplin	Blouses, dresses
Jersey	Knitted T-shirts, leisure wear
Felt	Crafts, hats, snooker tables
Tweed	Jackets, coats
Drill	Heavy-duty workwear
Organza	Bridal, high fashion

Standard components

Table 11.9 Standard components

Component	Types available include:
Threads Types of thread will be available in different weights, strengths and textures for different uses	Machine thread, tacking thread, top stitching, embroidery thread, monofilament

> **Exam tip**
>
> Know the common names of textile fabrics such as corduroy, chiffon or cotton poplin and not just their fibre source. In an examination it demonstrates a higher level of understanding. Cotton, for example, is the name of the fibre – there are many different variations when made into fabric.

Component	Types available include:
Fastenings Selected according to their function and intended purpose as well as aesthetic qualities	Buttons, poppers, zips, velcro, hooks and eyes, toggles
Structural components These are used to add support and help shape a product	Boning, Petersham
Other components These components provide functional support or are chosen for aesthetic reasons	Bias binding, elastic, ribbons, laces, beads, sequins, eyelets

Figure 11.12 An assortment of textile components

Cost and quantities

The cost of raw materials and components should be considered when making a textile product. The following are points to consider:

- the cost of the fabric and the width which will give the best use of fabric
- the fibre type can affect the overall price
- all component parts need to be costed
- bulk buying of fabric and components in industry can substantially reduce the overall cost of raw materials
- pattern templates must be laid correctly to minimise the amount of waste fabric.

> **Typical mistake**
>
> In maths questions where calculations are needed, such as working out costs, make sure you show all of your workings even if you have used a calculator. You will lose marks if you do not show the method used.

Manufacturing to different scales of production

REVISED

One-off, bespoke or job production

- **One-off**, custom-made or **bespoke** products are made by an individual or a small team of highly skilled, versatile workers with the ability to adapt to a range of processes and machinery.
- One-off products are often commissioned as unique pieces by clients who may also have direct input into the design process.
- They often use high quality fabrics and components and are therefore expensive to purchase.
- Many haute couture, designer gowns are made in this way.

> **One-off**: a single product.
>
> **Bespoke**: made to measure for an individual client.

Batch production

- **Batch production** is used to produce a specific number of identical products in a set timescale such as seasonal fashion products and are usually of mid to low quality.
- The products are made by large teams of workers who specialise in one element of the construction process, which means work can be repetitive and boring.
- Batch production is flexible and can change to meet market demand. Repeat orders can be facilitated quite easily.

> **Batch production**: a limited number in a set timescale.

Mass production

- **Mass production** is the largest scale of production, usually for products that are in high demand over a long timescale. Many factories run 24/7 to maximise output and profit.
- Typical mass-produced products include socks and plain t-shirts, where styles rarely change.
- Workers are skilled or specialised in one element of the construction process.
- CAM is increasingly used in mass production where consistency and speed are important.

Manufacturing systems

Mass and batch production lines can be organised in different formats in order to maximise efficiency and output.

- Straight line production: the work is passed from one worker to the next either along a conveyor belt or on overhead automated systems.
- Progressive bundles: bundles of garments or product parts are moved in sequence from one worker to the next. Each worker completes a single operation within the bundle before passing it on.
- Cell production: groups of workers operate together to make whole or part products. Cells operate separately within other systems.

> **Mass production** : Very large numbers made continuously over long periods of time.

> **Typical mistake**
>
> You may lose marks if descriptions of mass and batch production do not differentiate sufficiently between the two systems. Similar products can be made under both systems, but quantities and timescale will be considerably different.

> **Exam tip**
>
> When a question starts with 'explain', you must state a fact and provide further elaboration of that fact. You will lose marks if you simply list different facts.

Specialist techniques and processes

REVISED ☐

Tools and equipment

Hand tools for manufacturing textile products include:

- metre rule: used for marking out and cutting fabric and measuring hemlines
- tape measure: for taking body measurements
- craft knife: for cutting small templates and stencils
- cutting mat: includes guides for accurate cutting and are used with a rotary cutter or craft knife
- quick un-picker: to unpick faulty seams
- pins: a temporary method of holding fabric together.

Sewing machines

- Domestic sewing machines have a wide range of facilities and stitches to complete different processes.
- Computerised sewing machines can also embroider original designs.
- Domestic machines have optional feet or specialist attachments for specific processes such as attaching zips.
- Sewing machine needles vary and should be selected according to the fabrics being used.
- An overlock machine is used in industry to cut a straight edge on the material and over sew the edge to neaten the seam in one process.

Laser cutting

- Laser cutting is controlled by a CAD program.
- The design is drawn as a 2D image which the laser follows to accurately cut or etch the design.

- The laser strength and speed is set depending on the material to be cut.
- Laser cutters are used in industry to cut through multiple layers of fabric on automated systems.
- Disadvantages include:
  - not all fabrics can be cut as some melt and burn
  - the laser leaves an unsightly burnt brown edge on some fabrics.

Pattern marking and cutting

- Pattern templates include pattern markings which must be followed to ensure that the finished product is the correct size, shape and quality.
- The templates are placed on fabric and then cut around to create the pieces of fabric for the product.
- Fabrics can be laid out in different ways:
  - lengthways and folded over so that selvedge edges come together; the templates are then laid out according to pattern instructions
  - crosswise fold (folded along the width of the fabric) for a more efficient lay plan
  - on the bias, at a 45-degree angle to the straight grain, creating fabrics that are more flexible and can easily be sewn into curved shapes, but are wasteful.

Table 11.10 Pattern markings that must be followed accurately when making textile products

Pattern mark	Meaning of the mark	Why it is important
←————————→	Straight grain or grain lines	Template must be parallel to the selvedge edge, so the garment hangs as intended or lies flat
⌐————————⌐	Place on folded edge	The edge of the template needs to be on a folded edge of the fabric, as the piece is symmetrical
————————	Adjustment lines to lengthen and shorten templates	The templates can be adjusted here to get a more personalised fit
·-·-·-·-·-·-·-·	Cutting lines in various sizes	Cut along the desired size
··················	Stitching line	This is where stitches should be when joining pieces – if not adhered to, the product will not fit together
——✂—— - - - - - -	Seam allowance	The distance of the stitching line to the fabric edge, usually 1.5 cm
●	Dots	Indicates the position of a component or shaping technique
◆—◆◆	Notch	Indicate how pieces fit together and how to match a pattern, such as stripes
⊙	Position of button	Transfer the mark onto the fabric for correct placement on the garment
⊢————⊣	Position of button holes	Transfer the mark onto the fabric for correct placement on the garment
⋖	Position of dart	The dots need to match up to create the dart
⊏╤╤╤╤⊐	Placement of pleats and tucks	The lines need to match up to create the pleat or tuck

Tools used for transferring pattern markings to fabric include:

- tailor's chalk
- vanishing markers
- tracing wheel and carbon transfer
- tailor's tacks
- hot notch markers.

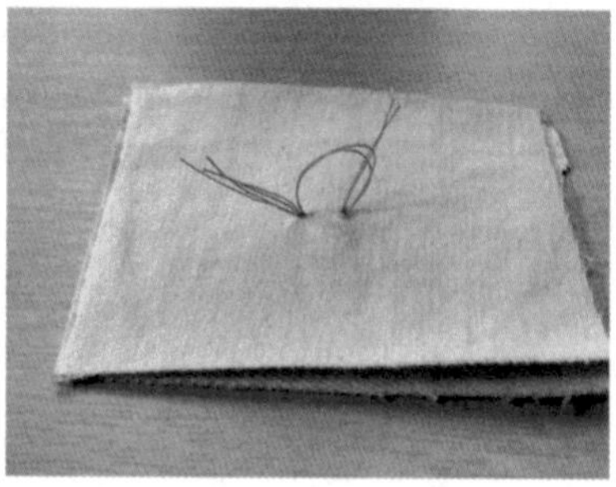

Cutting tools

Tools for cutting fabric include:

- fabric shears – these have long blades to easily cut fabric
- embroidery scissors – these have a sharp pointed blade for cutting intricate work
- pinking shears – these produce a cut zig-zag shape along the raw edge of a seam, which prevents fraying
- rotary cutters – sharp and accurate tools that are rolled along the surface of the fabric
- laser cutters – can cut complicated pattern pieces
- band saws – used in industry to cut through several layers of fabric accurately and quickly
- automated die cutters – used for cutting constant shapes through several layers of fabric.

Seam construction

It is important to use the correct type of seam when joining textile fabrics. The method used will depend on the fabric and product.

Tolerance and allowances

- The correct **seam allowances** and **tolerances** must be used throughout the manufacture of a product.
- The standard seam allowance used in textiles is 1.5 cm.
- If the correct seam allowance is not used consistently then product parts will not fit together as intended, leading to a poorer quality product.
- Some textile products are more complex so a permitted tolerance of a seam of about +/-1 cm is acceptable. However, this could still affect the overall size.

Figure 11.13 Tailor's tacks

Figure 11.14 Fabric shears

Table 11.11 Tolerances and allowances

Plain seam	French seam	Flat felled seam
Fabric right side / *Seam allowance* / *Fabric wrong side* / *Stitching*	*Second row of stitching* / *Fabric wrong side* / *Seam allowances* / *First row of stitching* / *Fabric right side*	*Fabric right side* / *Fabric wrong side*
• Most common seam type • Seam allowances can be stitched together or pressed flat open • Suitable for most fabric types	• French seams are enclosed and hide all **raw edges** • Used on more expensive clothing and sheer materials where seam allowances need to be hidden	• Strong seams with two rows of stitching adding to strength • Top stitching is often in a different colour • Lapped seams and double stitched seams look similar to flat fell seams

> **Seam allowance**: the distance between the raw edge of the fabric and stitching line.
>
> **Tolerance**: an allowance included in the seam allowance for inconsistency when assembling a product.
>
> **Raw edges**: fabric edges that are not neatened – unfinished.

Clipping seams

- Some curved seams need to be clipped along the seam allowance to allow fabrics to lie flat.
- This technique is particularly used around rounded areas such as necklines and armholes.

Finishing seams

- Seams are finished to prevent fraying. Methods include:
  - overlocking – excess fabric along the seam is trimmed off, the stitch overlaps the edge
  - zig-zag stitch – sewn along the raw edge of a seam
  - pinking shears – cuts a zig-zag finish and prevents fraying
  - seams and fabric edges can be bound using a bias binding.

Figure 11.15 Edge finished with a bias binding

> **Exam tip**
>
> Know how and where different seams construction methods are needed in the construction of a broad range of textile products. Be able to apply this in an exam situation.

Methods for adding body and shape

Table 11.12 Methods for adding body and shape

Method	Technique
Pleats	Fabric is folded back on itself and sewn in place, narrowing the original width of the material but adding shape Can be used as a decorative frill on products
Tucks	Similar to pleats and draw in fullness
Gathers	The fabric edge is gently drawn in to reduce and narrow the original width of the fabric, giving fullness and shape to a product Gathers can also be used as a decorative feature
Darts	Darts are made by creating folds in the fabric that taper to a point to improve shape and fit Darts are commonly used around the bust area but can be used anywhere to give shape

Figure 11.16 The pleats are stitched close to the waistband and lie flat but give shape to the skirt

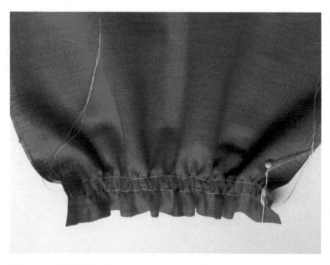

Figure 11.17 Gathers are formed by sewing two rows of stitching and pulling up the thread ends to create gathers

Edge finishes

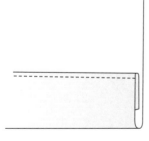

A basic double-folded hemline with a machine stitch

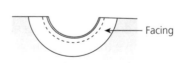
Facing

The facing is stitched in place and the seam is then clipped before the facing is turned to the inside

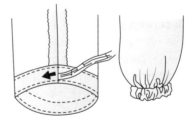

The elastic is inserted into a casing to form a cuff

Figure 11.18 Different methods of finishing edges

Computer-aided design and manufacture

Computer-aided design (CAD)

- Designers use CAD to develop surface designs for printing that can be transferred to relevant digital printing systems.
- Designs can be revisited, manipulated and changed, including the development of patterns and colourways.
- CAD is a more cost-effective way of developing designs. By using 3D imagery or virtual prototyping the need to make numerous prototypes is reduced. It is considered a more sustainable way of designing.
- CAD software is used to develop digital **lay plans**, allowing manufacturers to **tessellate** the pattern pieces to maximise fabric usage.

Computer-aided manufacture (CAM)

- Computers can be used to control sections of manufacturing in the textile industry. Some machinery is semi-automated, requiring some human input, while other machines are fully automated.
- CAM systems are expensive but speed up manufacturing, improve productivity and consistency and reduce human error.
- CAM applications include multi-head embroidery machines, digital printing on fabrics, laser cutters, 3D printers and automated fabric spreaders.

> **Lay plan**: how templates are laid out on fabric.
>
> **Tessellate**: how pattern templates fit together to use the least amount of fabric.

> **Typical mistake**
>
> Simple statements relating to advantages of using CAD such as 'it is quicker' will not get credit. Quicker than what? You could state it is quicker than drawing everything by hand which is more time consuming. Fully explain your answers!

Surface treatment and finishes

REVISED

Surface decoration techniques are used to improve the aesthetics of fabric, by adding colour, texture and pattern, for example, printing, dyeing, painting and embroidery. This fabric is then made into products.

Printing

Types of printing include:

- **Flatbed screen printing** – screens are used, applying a different colour and design to the fabric as it moves along a conveyor belt. When the design is complete the fabric will be fixed, washed and pressed.
- **Rotary screen/roller printing** – similar to flatbed printing except cylinders are used instead of screens. The fabric passes underneath the spinning cylinders, printing a continuous pattern onto the surface of the fabric. This is also referred to as roller printing.
- **Silk screen printing** – a fine mesh fabric is stretched over a wooden frame with part of the screen masked out with an opaque paste. The screen is laid face down on the fabric, printing ink is placed on the underside of the frame and a squeegee is used to drag the ink across the screen, forcing the ink through the fabric to leave the design on the material.
- **Stencilling** – a stencil is usually made from a thin sheet of card or plastic with a pattern cut out of it. Paint or dye is then applied through the holes in the stencil to leave a design on the fabric. Stencils can be hand cut or cut by a laser cutter.

- **Block printing** is a traditional method of printing. Paint or dye is applied to a relief block which is then pressed onto fabric to create a pattern. The process is repeated to create an overall repeat design.
- **Digital fabric printing** – large inkjet printers and specialist dyes transfer a digital image to the surface of the fabric. Can be used to create intricate and detailed images and to test sample pieces before production.
- **Discharge printing** – a bleaching agent is applied to fabric in the required design. The bleaching agent destroys the colour leaving a white or pale design.
- **Burn-out printing** – chemicals, usually acid, are applied to polyester/cotton fabric in the required design. The acid destroys the cotton fibre but the polyester resists leaving a lighter thinner section on the fabric.

Dyeing

- Dyeing is the most common method of colouring fibres and fabrics.
- Natural dyes work well on natural fibres, while synthetic dyes will give deeper or brighter colours.
- Synthetic fibres need chemicals to take on the dyes.
- Methods of dyeing textile fabrics include:
  - Piece dying: an entire length of fabric is dyed.
  - Dip dyeing: produces a graduated colour effect. Part of the fabric is dipped in the dye and is then gradually removed from the dye bath.
  - Tie and dye: the fabric is tied or knotted in a variety of ways to produce unique and varied designs. This is called a **resist method** of colouring.
  - Batik: another resist method, where hot melted wax is applied to fabric in the desired pattern. Once cooled the fabric is submerged in a dye bath or the dye can be directly painted on.
  - Random: dyeing or colouring small sections of a piece of fabric or yarns, different sections have different colours with no regularity to the design.

> **Resist method**: a means of preventing dye or paint from penetrating an area on the fabric. This creates the patterns.

Painting

- Fabric paints, fabric felt pens and pastel crayons are applied directly onto textile fabrics to create the desired design. Dimensional paints are also applied directly to fabric to give a slightly raised 3D surface.
- Specialist silk paints give a very delicate watery effect and can be used with Gutta outliner, which acts as a barrier to separate sections of the design.
- Marbling paints, which can be swirled into a design, are floated on a specialist paste. The design is transferred to the fabric when the pre-treated fabric is laid carefully onto the paint and paste.
- Air brushing requires a specialist air-operated tool that sprays dye or paint onto fabric, creating a smooth blended effect.

Figure 11.19 A hand painted silk scarf which uses a Gutta outline to create a barrier between the different sections

Embroidery

Embroidery designs vary depending on whether they are prepared by hand or by machine. Types of machine embroidery include:

- Free machine embroidery: the fabric is secured in a frame and the machinist moves the fabric around freely to create a design.
- Machine embroidery: in-built decorative stitches which can be used to enhance any designs.
- Computerised sewing machines: either have pre-installed designs or are linked to CAD packages to create and then stitch original designs (CAM).
- Appliqué: a traditional way of applying a design to fabric by stitching separate pieces of fabrics onto a base fabric. There are endless possibilities for creative design work through mixing colour, textures, stitch type and number of pieces.
- Embroidery: often further enhanced with beads and sequins.
- Beadwork: beads are used to enhance other techniques such as appliqué and can be placed randomly, used to outline an area or in a cluster.
- Beads are available in many different shapes, sizes and material. Similar effects can be achieved using sequins or diamantes.

> **Typical mistake**
>
> Show understanding of the stages needed to apply surface treatments and finishes to fabrics. Stages in a process need detail to fully explain the stage. Marks may not be awarded for a sequence that lacks detail, is incomplete or in an incorrect order.

> **Exam tip**
>
> 'Evaluate' when used as a command word in exam questions requires evidence of appraisal or making judgements. Statements made should reference both positive and negative viewpoints.

Now test yourself

TESTED

1. Explain why a satin weave fabric is less stable than a twill weave fabric. [3]
2. Give three reasons why the intense farming of cotton crops is detrimental to the ecosystem. [3]
3. Explain the importance of following pattern marking when laying out templates on fabric. [2]
4. Explain the purpose of a cell operating within a manufacturing system. [3]
5. Outline the reasons for using different seam construction methods in product manufacture. [4]

Exam practice questions

1. An outdoor security light switches on for 45 second when movement is detected.
   (a) From the list below, circle the subsystem which will produce the 45-second timer function for the security light:

   logic gate   amplifier   monostable   astable   comparator [1]

   (b) The timer subsystem uses a resistor and a capacitor. The formula for the time delay is: T = 1.1RC

   T is in seconds

   R is the resistor value in ohms (Ω)

   C is the capacitor value in farads (F)

   The timer uses a 220µF capacitor.

   Calculate the resistor value to produce a 45s time period. [3]

(c) The outdoor security light is housed inside a box made from mild steel sheet.

(i) Describe **two** problems a designer must consider when developing a security light for outdoor use. [2]

(ii) Use sketches and notes to describe a manufacturing and finishing method for producing a prototype box for an outdoor security light from mild steel sheet. Identify tools, processes and materials used. [6]

(d) Discuss ways in which an outdoor security light can be developed as part of a sustainable design strategy. [4]

2 Complete the sectional views through the different papers and boards below.

| Corrugated card | Foam board | Corriflute |

3 Explain the advantages of using manufactured boards rather than natural timber. [4]

4 Describe the life cycle of a soft drinks can made from aluminium. [6]

5 Study the picture of the rainwater guttering.

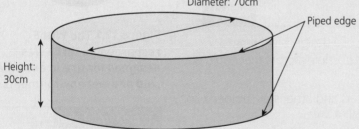

(a) Name a suitable thermoforming polymer for the rainwater guttering. [1]

(b) The rainwater guttering has been made by extrusion. Explain why extrusion is a suitable production method for the manufacture of rainwater guttering. [3]

6 Components are used in textile products for a range of different reasons.

(a) Give **two** reasons for using piping as an edge finish on a textile product. [2]

(b) A textile student has decided to make a circular floor cushion that will have a piped edge inserted into the seams of the two circular end pieces, as shown on the diagram below.

Diameter: 70cm

Piped edge

Height: 30cm

(i) Calculate how much piping cord will be needed to complete the cushion. [4]

(ii) The sides of the floor cushion will be cut from one rectangular piece of fabric.

(c) Calculate the size of the rectangular template that will be needed to form the side of the floor cushion. Include seam allowances of 1.5 cm in your calculation. [2]

ONLINE

3 Core designing and making principles

12 Understanding design and technology practice

Context of a design

- The **context** of a design incorporates many things such as:
  - the surroundings or environment where it will be used
  - the different **users** and **stakeholders**
  - the purpose of the end product
  - social, cultural, moral and environmental considerations.
- A product that is designed properly within context will fulfil its purpose exactly, giving the users and stakeholders what they require with a minimum amount of interaction or inconvenience.
- If design takes place without consideration of the context it will result in a final product that does not fully meet the needs of the user or stakeholders.

> **Context**: the settings or surroundings in which the final product will be used.
>
> **User**: the person or group of people a product is designed for.
>
> **Stakeholder**: a person other than the main user who comes into contact with or has an interest in the product.

Figure 12.1 The WOLF rechargeable torch was designed for use in extremely cold and wet environments

Context map/task analysis

A useful method of considering the context of a design is to create a context map or task analysis showing all the possible factors that could or should influence the design, for example:

- Who are the primary users, secondary users and other stakeholders? For example, their age, gender, physical mobility.
- Why is there a need for this product? What is the problem? What are the restrictions on the design?
- Where will the product be used? What type of environment will it be used in? For example, is it for use indoors or outdoors?
- What must the product do? What is its primary function? Does it have just one main purpose or must it fulfil a number of different requirements?

> **Typical mistake**
>
> Although the user may often appear to be the only stakeholder, this is rarely the case. Anyone who may come into contact with a finished product can be considered a stakeholder.

- When is the product used? Is the product used at certain times of day/night? Is it used at certain times of the year then stored for long periods of time between uses?
- How is the product meant to function? How will it be stored, transported, maintained? For example, will it have to function without making any noise?

> **Exam tip**
>
> Toys are designed for children, but parents purchase the toy and have to store, carry, clean and maintain it. Although the child is the main user, the parents are a major stakeholder and their needs must also be considered by the designer.

13 Understanding user needs

Needs and wants of the end user
REVISED ☐

- One of the main considerations for any design must be the user group and their needs.

The main person or group of people who will be using the product are known as the **primary users**.

- Assessing the user and stakeholder needs will require data to be collected.

> **Primary user**: the main user of the product.

Primary data
REVISED ☐

- **Primary data** is data collected first-hand from the main user or stakeholders. For example, using questionnaires, surveys, interviews or studies you conduct yourself.
- If a product is to be marketed and aimed at a wide user group then hundreds of questionnaires, interviews and tests across the spectrum of different users will need to be conducted.
- Failure to carry out sufficient research can lead to inaccurate primary data and a product that does not fulfil the user's or stakeholder's needs.

> **Primary data**: research collected 'first-hand' by yourself.

Collecting primary data

- Surveys and questionnaires are an effective way of gathering information from people, providing the questions are well thought out and worded.
- Questions can be open or multiple choice depending on the type of information you require.
- Limiting the answers to a choice of three or four options means that you won't get irrelevant answers.
- Finding people within your intended user groups to fill out surveys is of prime importance but can be difficult.

Figure 13.1 Interviewing and asking the primary user's opinions

- People who fill out your survey or questionnaire must fall into the same user group as your intended users or the information collected will be inaccurate and result in a flawed final product.
- Many free online survey programs are now available, meaning you can email people who fit into your intended user group.

Secondary data

- **Secondary data** is 'second hand' data which has already been collected by someone else.
- It is usually much easier to find than primary data, but it may already be out of date.
- Secondary data can be gathered from a range of sources such as websites, books, test reports and journals.
- Using secondary data can save time as it is much quicker than carrying out testing, interviews and questionnaires and is therefore much less expensive.
- Data collected is not as accurate as primary data because it is not specific to the designer's or user's exact needs.

> **Secondary data**: research collected by others.

Collecting secondary data

- Images and information on existing products from the internet are based on the reviews of others and/or the manufacturer's own claims.
- It is better to find a 'real life' existing product which you can handle, use and test yourself.

Now test yourself

TESTED

1 State three ways of collecting primary data. [3]
2 Explain the advantages and disadvantages of using secondary data for research. [4]

14 Writing a design brief and specifications

Design briefs

REVISED

- A design brief is a concise description of the task that the designer will undertake to solve the design problem or achieve what the **client** wants.
- The design brief can be set by the client or decided between discussion by the client and designer.
- It should be short and give a clear outline of what the required results of the design are.
- The designer must refer to the brief during the design process to ensure they are working towards achieving this.

> **Client**: the person the designer is working for (this may or may not be the user).
>
> **Design fixation**: when a designer limits their creativity by only exploring one avenue of design or relying too heavily on features of existing designs.

Writing a design brief

- Before starting it is important that the full extent of the problem and context has been fully explored.

- What a client wants and what they might actually need are not always the same thing.
- When the brief focuses more on the product than the problem this can lead to **design fixation** and prevent alternative approaches being considered.

Specifications

REVISED

- The specification is a set of requirements that the product must meet or constraints that it must fit into.
- Research carried out into the problem should also influence what goes into the specification.
- **Open specifications** state **criteria** that the product must meet but do not specify how it must be achieved.
- **Closed specifications** are more detailed and state what must be achieved and how certain criteria must be met. For example, by specifying the tools, materials or processes that must be used.

Writing a specification

The specification will usually cover areas such as:

- the primary and secondary functions that the product must achieve (what it must do)
- any specific requirements of the user/stakeholders
- materials and components that must be used or avoided
- maximum or minimum dimensions, weights and size constraints
- financial constraints (how much it should cost to produce)
- aesthetic factors (how the product must look or feel)
- anthropometric and ergonomic requirements
- environmental standards or constraints
- safety features and restrictions
- relevant manufacturing standards
- legal requirements
- how long it should be expected to last for.

Various websites and software packages are available to assist when writing a specification by providing prompts or questions about the product they are intending to design. These include ACCESS FM and SCAMPER.

> **Open specification:** list of criteria the product must meet but not specifying how it must be achieved.
>
> **Criteria:** specific goals that a product must achieve in order to be successful.
>
> **Closed specification:** list of criteria stating what must be achieved and how it must be met.

Now test yourself

TESTED

1 What are the key differences between the brief and the specification? [2]
2 What can help prevent design fixation? [2]

15 Investigating environmental, social and economic challenges

Environmental considerations

REVISED

- Until recently, manufacturers produced products as cheaply and quickly as possible with little concern about the environmental impact.
- Social pressures to have the newest products and the influence of fashion and trends mean people throw away rather than repair things. This is known as a **throwaway society**.
- The throwaway society follows a linear economy – see Topic 1.
- A circular economy uses as few resources as possible and extracts the maximum from them, using them for as long as possible – see Topic 1.

Figure 15.1 The use of plastic when designing products can have harmful effects on the environment

Designing environmentally friendly products

Eco design is designing sustainable products that will not harm the environment by, for example:

- choosing materials which are sustainable, recyclable and non-toxic and do not require as much energy to process
- designing products that:
  - are fuel and material efficient
  - last as long as possible so there is less replacement of parts
  - function to the best of their potential
  - can be fully recycled
  - use local materials, resources and labour.

A simple set of rules that cover the above are the Six Rs of sustainability – see Topic 1.

> **Throwaway society**: a society that excessively consumes and wastes resources.
>
> **Eco design**: designing sustainable products that will not harm the environment by considering the effects of the technology, processes and materials used.

Social challenges

REVISED

Positive and negative impacts of products

- Products can have a negative impact on society that the designer may not have foreseen.
- While it is not always easy to predict what the effects of a product might be, designers must try to consider the possible negative implications and decide whether the product is worthwhile.

Cultural awareness

- **Cultural awareness** is considering how different cultural groups may be affected by a product. Different cultures may view products differently depending on their beliefs, ideas and experience.
- A product designed for a particular market sector (such as the UK market) should not be offensive or upsetting to anyone.
- Designing new products that continue traditional processes, skills and materials can help preserve cultural identity.

> **Cultural awareness**: understanding the differences in attitudes and values between people from other countries or other backgrounds.

Now test yourself answers at www.hoddereducation.co.uk/myrevisionnotes

Economic challenges

- The manufacturing processes, transportation, use and disposal of a product can affect the economic needs of people.
- The manufacture of a new product may create jobs, boosting the overall economy of the area.
- If the product is manufactured by modern CAM machines it could lead to the loss of jobs and in turn affect the economy in that area.
- Many products are made abroad because labour costs are much cheaper. Some workers in these countries are poorly paid, work in dangerous or unhealthy conditions and work long hours without breaks or holidays.
- In some countries there is use of child labour.
- Designers should make sure their clients follow the guidelines for fair treatment of workers set out by organisations such as the European Institute for Crime Prevention and Control, the GLA (Gangmasters Licensing Agency) and AWARE (The Alliance for Workers Against Repression Everywhere).
- Manufacturers of Fairtrade products ensure that their workers are treated fairly and are not **exploited**. For more on Fairtrade see Topic 1.

> **Exploitation**: the act of treating someone unfairly in order to benefit from their work.

Anthropometrical and ergonomic issues

- **Anthropometrics** uses statistics and measurements of different parts of the human body taken from people of different ages, genders and races.
- This is used to determine the sizes, shape and forms of products so that they meet the needs of the intended user.
- In some cases, products are designed to fit the average person, in others they may be designed to fit the largest or smallest.
- **Ergonomics** is the relationship between people and the products which they use.
- Ergonomic data is used to make products more comfortable and easy to use by considering the shape of the human body, and the force a person can apply to something.
- Ergonomics and anthropometrics are used together when designing products to ensure the outcome not only 'fits' the intended user but is as comfortable and easy to use as possible.

> **Anthropometrics**: the study of the sizes of people in relation to products.
>
> **Ergonomics**: the relationship between people and the products which they use.

Now test yourself

TESTED

1 What is the difference between anthropometrics and ergonomics? [2]
2 Describe how mobile phones have had both a positive and negative impact on our society. [8]

16 Developing ideas

Testing and evaluating ideas

- Innovative design is exploring new ways of doing things or trying to approach problems from different angles or perspectives.
- To get different perspectives on a problem in order to explore new ideas, designers can use different sources of information.
  - Focus groups – a group made up of different people who each give their views and experiences of a problem or discuss their perceptions and attitudes towards an existing product or idea.
  - Existing solutions – looking at existing solutions to a problem or products designed to solve similar problems and considering which aspects work well or need improvement.
  - Biomimicry – looking at different aspects of the natural world such as naturally occurring structures and features, then considering how these methods, shapes or forms could be incorporated into the design (for more on biomimicry see Topic 2).
- **Development** is the process of selecting ideas, elements, materials and manufacturing techniques from initial ideas and using them to explore and produce better designs or ideas.
- Development is the part of the design process where designers try out new things and make creative decisions based on what works and what doesn't.
- **Modelling** allows designers to try out and test ideas or parts of designs by making scale models. Modelling and testing allows designers to see if a design will work and then make further decisions or developments.
- Models and ideas that do not work are just as valid as often more can be learned from a failed or unsuccessful model.
- By taking risks and trying new things, designers can come up with innovative ideas. Each step along the way is a vital part of the development process.

> **Development**: the creative process of selecting ideas, elements, materials and manufacturing techniques from initial ideas and using them in new ways to explore and produce newer and better designs or ideas.
>
> **Modelling**: trying out and testing ideas or parts of designs by making scale models.

Critical analysis and evaluation

- Critical analysis and evaluation go hand in hand with development but can be done at any point in the design process.
- Critical analysis will assess the suitability of the design against given criteria to check it meets the necessary requirements set out.
- The list of criteria will depend on the product but will cover:
  - **Product function**
    - Does the product do what it is supposed to?
    - How well does it do this?
    - Is it easy to use?
    - Is it comfortable to use?
  - **Aesthetics of the product**
    - Does it look appealing?
    - Does it feel nice?

- ○ **Anthropometrics and ergonomics**
  - – Is it the correct height, length, diameter?
  - – Does it suit all users?
  - – Can it be adjusted to suit different users?
  - – Will it fit where it is supposed to?
- ○ **Cost**
  - – Will the materials be expensive to purchase?
  - – Will the manufacturing costs be acceptable?
  - – Will the final price of the product be acceptable?
- ○ **Materials**
  - – Are the materials used of high quality?
  - – Are the materials suitable for the product?
  - – Are the materials easy to source/readily available?
- ○ **Construction methods**
  - – Is it well made?
  - – Are the skills/processes required to make it readily available?
  - – Is it quick to manufacture?
- ○ **Health and safety**
  - – Is it safe to use?
  - – Does it meet relevant health and safety standards?
- ○ **Environmental considerations**
  - – Are the materials from a sustainable source?
  - – Can the product be easily disassembled?
  - – Can the materials be recycled?
  - – Are the processes used harmful to the environment?

Refinement and modification

REVISED ☐

- ● Following the critical analysis, the designer can make further changes or modifications in order to address any areas where the design is not meeting the criteria or not performing well.
- ● The modified design is then re-analysed and if further improvements are needed the design will be modified and the same process repeated.

This is called '**iterative design**'.

- ● Each new 'iteration' of the design should be a slightly improved version of the last until the final iteration which should meet all the criteria as much as possible and is the best possible solution.

> **Iterative design**: a repeated cycle of quickly implementing designs or prototypes, gathering feedback and refining the design.

Now test yourself

TESTED ☐

1 State four criteria that would be considered when critically analysing a product. [4]
2 Describe an example of how a designer might use biomimicry. [3]

17 Investigating the work of others

Airbus

- Airbus is a European company best known for aircraft, but which also has divisions that focus on helicopters, military equipment and space travel.
- The Airbus A380 is the world's largest passenger aircraft carrying up to 800 passengers.
- Engineers overcame the issues around the size and weight of the A380 aircraft by:
  - using lightweight composite materials to reduce the overall weight
  - looking at biomimicry for inspiration, in particular the shape and structure of eagle wings to overcome issues with the wingspan.
- **Generative design** is used to optimise components or parts to reduce weight but maintain strength.
- Airbus use CAD technology throughout development, including 3D printing of parts from a variety of materials, particularly titanium, because of its excellent strength-to-weight ratio.
- Airbus is a global company – wings are manufactured in Britain, the rear fuselage in France and the front fuselage in Germany.
- The Airbus Beluga is the company's super transporter as transportation of parts is vital to the company's success. The huge capacity of the Beluga reduces the environmental impact as fewer journeys are needed to transport parts.

Figure 17.1 Airbus A380

Generative design: a computer-based iterative design process that generates a number of possibilities that meet certain constraints, including potential designs that would not previously have been thought of.

Apple

- Much of Apple's success is due to the importance they place on innovation and design.
- Apple were pioneers in the use of graphical user interfaces (GUIs).
- The Apple Lisa computer in 1983 was the first desktop computer to have icons or small images that represent files, folders and discs and a cursor controlled by a mouse. The cursor and mouse system is still used today despite the introduction of touch-screen technology.
- Apple products are easily distinguishable by their sleek design, consistent shapes, colours and materials.
- British industrial designer Jonathon Ives, appointed in the mid-1990s, was responsible for the styling of the first iPod and iMac. Aesthetics and user experience were put at the forefront of Apple's design philosophy.
- Apple are often criticised for the products being developed with planned **obsolescence**. They are also criticised for software updates that do not work on older products and for developing their own ports for connecting other devices.

Obsolescence: when a product is out of date or no longer useable.

James Dyson

REVISED

- James Dyson is known for creating innovative products that use new technology and engineering principles and improve on existing products.

- One of his early products was the ball barrow, a variation on the traditional wheelbarrow. The ball barrow spread the load, making it easier to push on soft ground and improve manoeuvrability.

- Previously, vacuum cleaners relied on bags to house dust that had been collected. Dyson noted the fuller the bags became, the more suction levels dropped making the vacuum cleaner less effective. The DC01 vacuum cleaner used cyclonic technology to collect dust without the need for a bag.

- Dyson went thorough 5127 iterations of the DC01 product over ten years before gaining market interest.

- Other products developed under the Dyson name include washing machines, fans, hand dryers and hairdryers. All go through rigorous and extensive testing before the products are launched.

Figure 17.2 Dyson ball barrow

> **Typical mistake**
>
> Avoid writing about personal or biographical details relating to the designers you have studied. This will not gain any marks in an exam. Questions will test what you know and understand about their work – their design thinking, the inspiration behind their work and any influence they may have on product design and manufacturing.

Philippe Starck

REVISED

- Philippe Starck is a well-known and influential designer perhaps best known for his commissioned work with Alessi, the Italian homeware manufacturer, and the lighting company FLOS.

- His most well-known product is the elaborate lemon squeezer known as 'Juicy Salif' which was inspired by the shape of a squid. He is reported to have said this design is not meant to juice lemons but start conversations.

- Starck is known for using pioneering manufacturing techniques and materials not commonly used for the type of product, for example, the one-piece injection moulded polycarbonate chair – the 'Louis Ghost'. This chair has no other fixings and the transparent material is most commonly used in low-cost disposable products.

- Starck is also known for designing buildings, interiors, clothing, luggage and watches.

- One of his most controversial building designs is the Asahi beer hall in Tokyo that features a large golden flame on its roof.

Matthew Williamson

REVISED

- Matthew Williamson's signature style includes:
  - the strong use of vibrant jewel-like colours such as fuchsia, orange and turquoise
  - printed patterns which include dragonflies, butterflies, peacock feathers and exotic plants
  - heavy **embellishments** such as embroidery and beading
  - fabrics such as chiffon and silk.
- His signature style has had a lasting impact on the high street, particularly in his use of colour, pattern and embellishments such as embroidery to add texture.
- Travel to exotic locations such as Morocco or Africa is often the inspiration behind his work.
- His vision is to make women feel beautiful and feminine, and to enjoy wearing his clothes.
- His signature style and influence can now be seen throughout contemporary fashion, including interior design.

> **Embellishments**: surface decoration such as embroidery and beading.

Figure 17.3 'Electric Angels' was Matthew Williamson's first collection in 1997 and consisted of 14 pieces made in vibrant colours

> **Exam tip**
>
> Read exam questions carefully to make sure you fully understand what the question is asking before attempting the answer. Underline or highlight key words to help focus on what is important.

Now test yourself

TESTED

1 Explain how biomimicry inspired engineers when developing the Airbus A380. [2]
2 Describe the features that make Apple products easily recognisable. [3]
3 Describe the impact Dyson's DC01 vacuum cleaner has had on the design of other vacuum cleaners. [3]
4 Explain what was innovative about Alessi's 'Louis Ghost' chair. [4]
5 Describe three features that represent Matthew Williamson's signature style. [3]

18 Using design strategies

Collaboration

- Many designers work in **collaboration**, in pairs or in groups. Usually a design business will employ a number of designers who work together on specific design projects.
- By discussing, sharing and working with each other they bounce ideas back and forth, refine elements or go down completely new avenues of investigation.
- Once a designer has created some initial ideas or concepts, a focus group comprising the client, user and main stakeholders give feedback on the designs.

> **Design collaboration:** a number of designers working together on specific design projects.

User-centered design

- **User-centered design** puts the user at the 'centre' of the design process.
- The user-centered design process has four main stages.
  - Specify the context of use: identify the users of the product, what they will use it for and under what conditions it will be used.
  - Specify requirements: identify any user goals that must be met for the product to be successful.
  - Create design solutions: this may be done in stages, building from a rough concept to a complete design.
  - Evaluate designs: evaluation through usability testing with actual users.

> **User-centered design (USD):** looking at and checking the needs, wants and requirements of the user at every stage of the design process.

Systems thinking

- **Systems thinking** is when you consider the product you are designing as part of a larger system or experience.
- The opening of the packaging, maintenance of the product, use of the product and disposal or exchange of the product are all part of the experience of owning a product.
- Systems thinking considers the whole problem and how to provide the best service to the user.

> **Systems thinking:** considering a design problem as a whole experience for the user.

Now test yourself

TESTED

1 Describe the benefits and possible drawbacks of collaboration when designing. [4]
2 Name the four main stages of 'user-centred design'. [4]

19 Communicating ideas

Formal and informal 2D and 3D drawing

- Two-dimensional drawings are useful for showing simple profiles (shapes), the layout of parts within a design, or for showing cross-sections through a design.
- Three-dimensional drawings are useful for showing design ideas.

Freehand sketching

- Sketches made without the use of templates, grids or other drawing aids.
- Feint lines may be sketched to 'crate out' simple block shapes which the designer will later refine into the required shape.
- Drawings can be 2D or any style of 3D.
- Brief annotation may be used.

Oblique drawing

- A basic, low-skilled, form of 3D drawing.
- A 2D profile is projected into a 3D shape by drawing lines at 45°.
- A 45° set square is usually used, for accuracy.

Isometric drawing

- A 3D drawing method that produces a more accurate representation of a shape.
- An angle of 30° is used for the projected lines.
- A 30° set square or isometric grid paper is often used for accuracy.
- All lines are drawn to their full length.
- Large isometric drawings can appear distorted in shape because they have no 'perspective'.

Perspective drawing

- A 3D drawing method that achieves a very realistic representation.
- In two-point perspective, two 'vanishing points' are drawn at either end of an imaginary horizon.
- The projected lines from the front edges of the drawing all move towards the vanishing points, causing them to converge.

> **Oblique drawing**: a basic 3D drawing method, using lines projected at 45°.
>
> **Isometric drawing**: a 3D drawing method, using 30° angles for projections of depth.

Figure 19.1 Oblique projection drawing

> **Exam tip**
>
> Questions may ask you to draw a 3D (isometric) projection of an object from a series of 2D (orthographic) drawings, or vice-versa.

> **Perspective drawing**: a realistic 3D drawing method.

Figure 19.2 Two-point perspective drawing

Thick and thin line technique

- A simple method of improving the impact of a 3D sketch.
- A thicker, bolder line (using a pen or a dark pencil) is used to 'go over' all edges of the drawing where the detail on an adjacent edge is not visible.

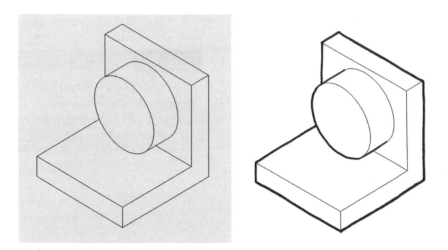

Figure 19.3 Thick and thin lines used on an isometric drawing

Colour and shadow

- A skilled technique of showing surface patterns, texture and depth.
- Essential for representing textiles and fabrics.
- Shadows and shading can show depth and shape of surface.
- Different **rendering** techniques can be used.
- 3D CAD programs can produce very realistic rendered drawings.

System and schematic diagrams

- Often used in electronic systems.
- A **system diagram** (or block diagram) is used to show the functional subsystems, how they are interconnected, and the signals flowing between them.
- System diagrams are broadly divided into input, process and output subsystems.
- A **schematic diagram** (often called a circuit diagram) indicates the connections between individual electronic components, and the values of the components.

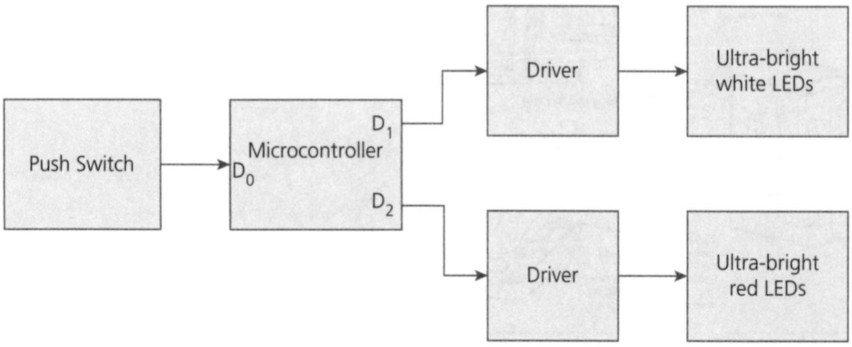

> **Rendering**: the method of using colour and shading to represent the nature of a surface.
>
> **System diagram**: shows the interconnections between subsystems in an electronic system.
>
> **Schematic diagram**: a circuit diagram showing the connections between individual components.

```
Push Switch → Microcontroller
            D_0
Microcontroller D_1 → Driver → Ultra-bright white LEDs
Microcontroller D_2 → Driver → Ultra-bright red LEDs
```

Figure 19.4 Electronic system diagram

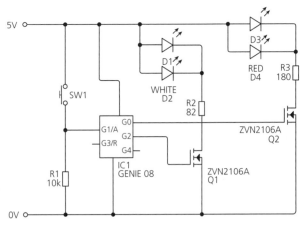

Figure 19.5 Electronic schematic diagram

Annotated sketches

Annotations should be used to add information which is not obvious, such as:

- arrows to indicate movement
- material used
- surface finish
- construction or manufacturing method
- function
- reference to commonly understood phrases, for example, 'headphone jack', or 'Velcro® attachment'
- hidden detail, for example, 'battery compartment underneath'
- reference to weight or balance
- samples of swatches or fabric for textiles.

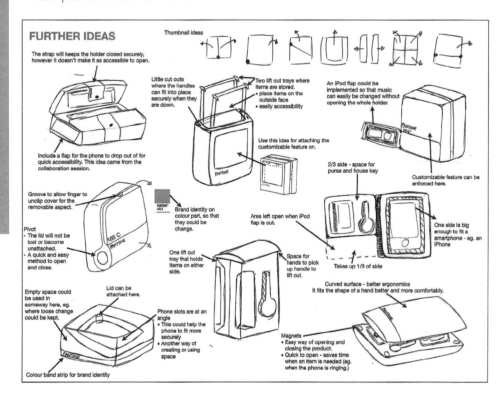

Figure 19.6 Annotated sketching

Exploded drawing

- A 3D drawing method used to show how components in an assembly fit together.
- Isometric drawing is usually used.
- The parts are separated along the 30° isometric lines, with arrows or dotted lines used to show how the parts fit back together.
- CAD programs can produce **exploded drawings** quite simply.

> **Exploded drawing**: a 3D drawing method to show how parts in an assembly fit together.

Models

REVISED

Models can be used to produce a full-size or scaled-down version of the entire concept, or they can be used to test a small part of the design, such as the function of a component or the finish on a material.

Cardboard models

- Cardboard is easily cut and low cost.
- Can be scored, bent and joined with a variety of adhesives.
- Rigid and available in various thicknesses.
- Can be laser cut from a CAD drawing.
- Useful for testing 2D linkages and mechanisms.

Foam models

- Foam core (or foam board) is similar to carboard, but more rigid. It is faced with white or black paper, giving it a very smooth surface, which can be rendered to give a realistic appearance.
- Polyurethane foam is used to create solid 3D models which can be interacted with – held in the user's hands. It is available in a variety of densities which can be easily cut and shaped using hand tools or CNC machines.
- Foam models can be finished and sprayed to look very realistic.

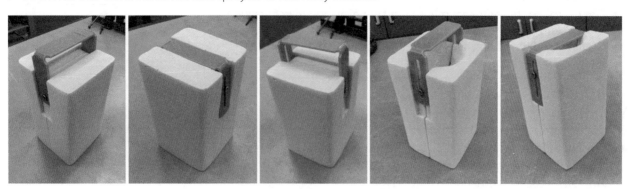

Figure 19.7 Using blue foam to model a retracting handle concept

Toiles

- A full-size textile model manufactured from a cheaper fabric.
- Allows the designer to check things such as garment fit or design and determine the position of zips and buttons.

Electronic circuits

Breadboards can be used to temporarily build a circuit for testing:

- no soldering is required
- they can be used to check component values, such as resistors
- they can be used to test the function of a microcontroller program.

> **Breadboard**: a temporary method of constructing electronic circuits for testing.

Presentations

REVISED

A designer may need to present design ideas to a third party. This could be done using:

- digital presentations, including slides, images, videos, sound and animations
- large-format presentation boards mounted on a display, showing sketches, diagrams and rendered images of the product.

Written notes

REVISED

- Used to formalise design thinking and to explain design decisions.
- Useful for explaining the creative process to a third party.

Flowcharts

REVISED

- A graphical representation of a process.
- Different symbols are used to represent different actions in the flowchart – see Topic 3, Figure 3.6.
- Flowcharts are often used to explain a process, such as a sequence of manufacturing steps.
- In a manufacturing flowchart, the decisions are often quality control checks.
- Flowcharts can be used to show a microcontroller program.

> **Exam tip**
>
> Questions may ask you to draw a flowchart to explain a manufacturing process (such as vacuum forming), including quality checks using decision boxes.

> **Typical mistakes**
>
> - Take care not to confuse a manufacturing process flowchart with a microcontroller program flowchart.
> - Understand the differences between a system diagram, a schematic diagram and a flowchart.

Working drawings

REVISED

- These are formal drawings that contain enough detailed information to allow a third party to accurately manufacture the design. They are sometimes called engineering drawings.
- They are often 2D orthographic drawings, showing a front, side and plan (top) view.
- They may include cross-sections or views of hidden detail.
- They contain information such as dimensions, scale, materials and details about component parameters (values).
- They may include a parts list.
- They should be drawn using British Standard symbols and methods.
- A textile working drawing is sometimes called a 'flat'.

Schedules

REVISED

- A diagram showing how a process can be completed within a specific timeframe in order to meet a deadline.
- A Gantt chart is a common way of showing a schedule.
- In industry, schedules can be used to plan when a particular machine will be needed, or to determine when materials need to be delivered.

Audio and visual recordings

REVISED

Can be used to record and present:

- findings from a focus group
- users interacting with a prototype design
- feedback and opinions about a model or a design
- evidence of the design functioning in its intended situation.

Mathematical modelling

REVISED

Especially useful when using CAD software. Mathematical modelling can:

- simulate the effect of forces applied to the design, for example, a structure
- test how a component responds to and conducts heat
- test designs in a virtual environment, saving time and expense.

Computer-based tools

REVISED

- CAD software is usually used to develop and refine initial ideas.
- Edits can be made quickly and it is easy to undo changes.
- 2D CAD designs can output to CNC machines, for example, a laser cutter.
- 3D CAD models can represent the colour and texture of the materials.
- CAD software can render the design to give it a photorealistic look.
- Common components, such as fasteners or hinges, can be selected from a parts library.

Now test yourself

TESTED

1 Name three types of 3D drawing techniques used by designers. [3]
2 Explain the difference between a system diagram and a schematic diagram in an electronic design. [3]
3 Give three benefits to a designer of making a model during the development of a design. [3]
4 Draw the flowchart symbols for:
   (a) input/output
   (b) decision
   (c) process. [3]
5 Describe two benefits to a designer of using CAD software. [4]

20 Developing a prototype

- Prototyping involves making a one-off version of the whole product or a specific part of the design.
- **Prototypes** can be used to test parts of the design, find out the users' views and identify any problems.
- Prototypes can show up potentially fatal flaws in a design which can be addressed before the product goes into full-scale production.

> **Prototype:** an early model of a product or part of a product to see how something will look or function.

Low-fidelity prototypes

REVISED

- **Low-fidelity prototypes** are produced early in the design process.
- They can be basic models of how a product will look or scale models of a part (such as a mechanism or feature) to illustrate or test how it works.
- Prototype models may be made from simpler materials such as paper or card instead of sheet metal and plasticine or modelling clay instead of polymers.
- Low-fidelity models are cheap and quick to make and allow designers to test crucial elements of their design and get quick results.

> **Low-fidelity prototype:** a quick prototype that gives a basic idea of a product's looks or functions.

High-fidelity prototypes

REVISED

- **High-fidelity prototypes** are made once a design has been developed considerably or a final design decided.
- They will look, feel and function as much like the finished product and be made using the same materials and processes as far as is possible.
- High-fidelity prototypes take much longer to produce and are more expensive but give a much more realistic idea of what the finished product will be like.

> **High-fidelity prototype:** a detailed and very accurate prototype similar to the final product.

When making prototypes to develop your ideas remember the following:

- Making something instead of just drawing it will help you to see your idea in a different way and find how you can improve it further.
- Don't spend a long time building a prototype as this will slow down the thought process and you will be less likely to change something if you have spent hours making it.
- Don't forget what the prototype is supposed to be testing and try to let the user test the product if possible.
- Don't be afraid to fail! If the prototype does not do what you want, use this knowledge to change or develop your design.

Figure 20.1 Low-fidelity prototype

Now test yourself

TESTED

1 Explain the term 'prototype'. [2]
2 Explain the difference between low-fidelity and high-fidelity prototypes. [4]

21 Making decisions

- Designers make decisions throughout the design process.
- Prototyping and asking for user feedback to test aspects of a design can help the designer make important decisions.
- The design brief and specification should be the reference points for all the decisions and the designer should constantly refer back to these.
- The most effective and useful feedback about a prototype is from the intended user. There are different ways in which feedback can be gathered.

User testing

REVISED ☐

User testing is watching a user interact and use your product for its intended purpose to see how well it works, how easy it is to use and whether they actually like it.

> **User testing:** testing by observing a user interact and use your product for its intended purpose.

Focus groups

REVISED ☐

- A **focus group** allows people to ask questions and state how they would like the product to be improved.
- This allows a wide range of responses to be gathered, but opinions may be different and contradictory because of the different users' needs.

> **Focus group:** group of people used to check a product design is on track.

A/B testing

REVISED ☐

- **A/B testing** is used to choose between two different design ideas.
- The results show which design achieved the task the quickest or most efficiently.
- This type of testing is often used to compare a new version of a product to the existing one to see if it works better.

> **A/B testing:** user testing to choose between two different design ideas.

Surveys and questionnaires

REVISED ☐

- Surveys and questionnaires are an easy way to gather information.
- The survey must ask questions that will provide accurate information.
- A questionnaire can reveal how well the product meets the users' needs, allowing the designer to make informed decisions about future modifications.
- The different types of feedback provide qualitative and quantitative data.
  - **Quantitative data** refers to data which gives specific counts and values in numerical terms, such as measurements like height, weight, size, humidity, speed and age. Data can be gathered from surveys, experiments or observations and presented in the form of charts, graphs and tables.
  - **Qualitative data** cannot be specifically measured but is observable by appearance, taste, feel, texture, gender and nationality. It can be collected from observations, focus groups, interviews and archive material. The data is presented as spoken or written words rather than numbers.

> **Quantitative data:** specific measurable data given in numerical form.
>
> **Qualitative data:** observations and opinions about a product.

All data gathered should be used by the designer to:

- evaluate the performance and suitability of the product
- make decisions about what needs to be improved
- implement the necessary modifications and changes to the design
- re-test and evaluate to check the effectiveness of the changes.

Now test yourself

TESTED

1 State three ways of presenting quantitative data. [3]
2 State two ways of gathering qualitative data. [2]

Exam practice questions

1 Testing and modelling are essential strategies in the iterative process of designing and developing new products.
Explain how Dyson exemplifies this process when designing and developing new products. [5]

2 (a) Give one advantage of using perspective drawing rather than isometric drawing when sketching design ideas. [1]

(b) Designers often produce three-dimensional models and CAD models when developing ideas. Describe one advantage of using three-dimensional models rather than CAD models. [2]

(c) Explain what is meant by a working drawing of a product. [3]

3 Designers have a responsibility to develop products which are as sustainable as possible. Explain three ways that designers can make products more sustainable. [6]

4 Many products are designed with planned or 'built in' obsolescence.

(a) Define the term 'planned obsolescence'. [2]

(b) Using an example, explain why products are manufactured with planned obsolescence. [4]

5 Symbols are often used on products or their packaging.
Name the symbol shown below and explain its meaning. [3]

ONLINE

Success in the examination

When will the exam be completed?

You will take the exam in the summer exam period of your final year, which will usually be in May or June.

How long will I have to complete the exam?

- The exam is two hours long.
- You should practise working on past papers and sample questions within the allotted time.

What type of questions will appear in the exam paper?

- **Section A** will consist of a variety of short answer, structured and extended writing questions that will test your knowledge of the core technical principles. You will have to answer all questions in this section.
- **Section B** consists of questions assessing your in-depth knowledge and understanding. You will have to answer one question in this section.

Tips on preparing for the exam

- If you did not understand a topic when it was covered in class, you are unlikely to understand it when revising. Make sure you ask at the end of a lesson if you are unsure of any of the material covered.
- Don't leave revision until the end of the course. Test yourself at the end of each topic.
- Use past papers, online materials and revision guides to help you practise exam-type questions.
- Plan your revision time in the weeks leading up to the exam.
- Make revision cards to help you compartmentalise your understanding.
- Work with other students to test each other.

Approaching the paper

- Make sure you know the date, time and location of your exam.
- Get a good night's sleep. Make sure you have eaten and that you are hydrated.
- Arrive early and make sure you have all your equipment with you.
- Read the instructions on the front cover of the question paper. This will tell you what you have to do.
- Read each question carefully at least twice. This will help you to understand exactly what information you need to give.

- The question will tell you how many marks are available for this question. Use this to gauge how much detail you need to put into your answer.
- If you finish early, go back and reread the questions and your answers. You will usually find that you have remembered more detail. You may also be able to spot any mistakes that you may have made.

Sample examination questions

Section A

Example

Study the pictures below which show children's toy shapes made from plastic (Toy A) and wood (Toy B).

Figures 22.1a and 22.1b
Children's toy shapes made from plastic (Toy A) and wood (Toy B)

(a) Complete the table below by stating which toy reflects the statement. [2]

Statement	Toy A or Toy B
The material used in this toy is chemical free.	
The material used in this toy comes from a finite source.	

(b) Discuss **two** environmental advantages that toy B has over toy A. [4]

ONLINE

Candidate response

(a)

Statement	Toy A or Toy B
The material used in this toy is chemical free.	Toy B
The material used in this toy comes from a finite source.	Toy A

(b) The wood used to make toy B is biodegradable, unlike the plastic in toy A. This means less of an environmental impact if it ends up in landfill as it will eventually decompose. Wood is also more long-lasting than plastic.

One mark for each correct response, in this case both answers are correct.

Each environmental advantage of wood needs to be fully discussed, including a comparison to plastic.

The first advantage is fully discussed and explained. The second advantage is correct but the fact has not been elaborated on. To gain the extra mark the candidate could have referred to plastic splitting more easily so the toy will not last as long and is likely to be thrown away into landfill much sooner than a wooden toy. Recycling the plastic could be an issue. Alternatively, wooden toys last longer and can be passed on through generations, unlike plastic toys, meaning less demand on new materials. Better in the long-term for the environment.

Now test yourself answers at www.hoddereducation.co.uk/myrevisionnotes

Section B

Sample question: Electronic systems, programmable components and mechanical devices

Electronic circuits can be constructed by different methods depending on the application.

Figure 22.2 shows a simple circuit built on prototype board. Figure 22.3 shows a printed circuit board (PCB), manufactured by batch production for use in an electronic product.

Study the photographs and describe three ways in which the printed circuit board has been designed to make it suitable for batch production, for use in an electronic product. [6]

Figure 22.2 Prototype board

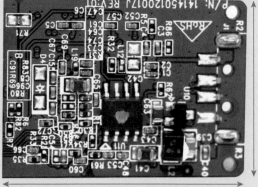

25 mm

40 mm

Figure 22.3 Printed circuit board

The first comment about soldering, while correct, has not been explained in terms of why this helps the PCB to be batch produced, and would not score any marks.

Candidate response

The components are so small that they can't be soldered by hand. They have been put on the PCB by a robot which works quickly and doesn't make mistakes so that means that a lot of PCBs can be made very quickly in a batch. The PCB has got a lot of components packed in a small space. This means you can make complicated circuits really small which is good for hand-held products. The white writing on the PCB tells the robot machine worker where to put all the parts so they don't waste any.

The comment relating to the components being put on by a robot is worth one mark, and the second mark is for explaining that this speeds up manufacture for batch production.

One mark is awarded for the comment about a lot of components being packed in a small space, and another mark for saying that this is good for hand-held products.

Better use of technical vocabulary (for example, 'surface mount technology', 'pick-and-place machines') would have improved the quality of this response.

The final comment about the white writing is incorrect and scores no marks.

Mark scheme

Three reasons must be given, with one mark for each valid point and one mark for explaining why that point makes the PCB suitable for batch production, or for use in a product.

Award credit for reference to any of the following points and any other valid points:

- Surface mount technology (SMT) allows the PCB to be manufactured by a robotic pick-and-place machine.
- Pick-and-place assembly is fast, accurate and reliable, so PCBs can be manufactured quickly.
- Soldering will be done in a reflow oven producing high-quality reliable joints.
- Miniature components allow high component density so the product can be made very small.

- The PCB will probably be double-sided (tracks on both sides) so complex circuits can be designed without tracks crossing.
- CAD will have been used to design and model the PCB, and this links with the CAM machines used to manufacture the PCB.

Sample question: Natural and manufactured timber

Study the picture of the spice rack. The spice rack is made from beech.

It is important that designers consider the world we live in and the needs of future generations. Evaluate how designers can lessen the impact on our environment when designing and making timber-based products such as the spice rack. [6]

Figure 22.4 **Spice rack**

The candidate has produced a coherent answer demonstrating relevant knowledge and understanding of how designers can lessen the impact on the environment when designing and making natural timber-based products such as the spice rack.

Candidate response

Designers should make the design as compact as possible, therefore using the least amount of material. They should make sure that the chosen timber is readily available and that it comes from a managed and sustainable source.

Designers should think about how the spice rack is to be manufactured and aim to reduce waste. They should aim to use renewable energy sources during production.

Designers should consider making the spice rack as a 'flat pack' product. This will reduce the negative environmental impact of packaging and transportation.

Designers should consider the type of finish to be applied to the spice rack. It should be durable to ensure the spice rack lasts as long as possible and also be water based so it does not harm the environment when cleaning brushes/spray guns.

The spice rack should be clearly labelled with recycling information to encourage its environmentally safe disposal at the end of its life.

The candidate has commented on the method of manufacture and the need to use power from a renewable source. They have suggested the use of the 'flat pack' method of assembly to minimise packaging and to assist in transportation.

They have included details regarding the reduction in the amount of material used and details of how to source materials from sustainable sources.

The candidate has discussed the use of environmentally-friendly finishes and the need to encourage recycling.

Mark scheme

Band	Descriptor
5–6	A coherent answer demonstrating detailed, relevant knowledge and understanding, to evaluate how designers can lessen the impact on the environment when designing and making timber-based products such as the spice rack. There will be evidence of relevant examples and well-developed substantiated judgements.
3–4	An answer with some structure, demonstrating partial knowledge and understanding, to evaluate how designers can lessen the impact on the environment when designing and making timber-based products such as the spice rack. There will be some evidence of relevant examples and partly substantiated judgements.
1–2	An answer demonstrating only basic knowledge and understanding, to evaluate how designers can lessen the impact on the environment when designing and making timber-based products such as the spice rack. There will be limited evidence of relevant examples or judgements.
0	No response

Sample question: Ferrous and non-ferrous metals

Study the picture of the centre punch.

The centre punch has been heat treated by hardening and tempering to ensure it is hard but not brittle.

Describe the processes of hardening and tempering.　[4]

Figure 22.5 Centre punch

Candidate response

To harden the centre punch, you would first clean the metal using emery cloth and then lay it on the firebricks in the brazing hearth.

You would then use the brazing torch to heat up the metal until it was cherry red.

Then you should quench the metal into cold water while it is still red. This would harden the punch.

To temper the centre punch, you would again clean it with emery cloth and lay it onto the fire bricks in the brazing hearth.

You would then heat the centre punch until it changes to a dark straw colour and leave it to cool.

The candidate has produced a full answer demonstrating relevant knowledge and understanding of the processes of hardening and tempering. The answer is logically sequenced and contains all the correct technical terminology.

Mark scheme

One mark is awarded for each correctly described operation of hardening and tempering up to a maximum of four marks. Both hardening and tempering must be addressed for full marks.

Hardening:

● clean the metal with a file or with an abrasive cloth
● heat the metal to red heat
● quench in water.

Tempering:

● clean the metal with a file or with an abrasive cloth

● heat the metal until a dark straw colour.

● allow to cool.

Sample question: Thermoforming and thermosetting polymers

Study the picture of the yoghurt pot.

The yoghurt pot has been made by vacuum forming.

Describe the processes of vacuum forming. [6]

Figure 22.6 **Yoghurt pot**

Candidate response

A yoghurt pot mould is placed on the table of the vacuum-forming machine.

The platen is then lowered into the machine and a sheet of thermoforming polymer is clamped over the mould. The thermoforming polymer is then heated until it becomes soft.

The platen is then raised and the vacuum switched on.

The thermoforming polymer is then sucked around the mould, the heat is switched off and the sheet is allowed to cool.

The candidate has produced a good answer worthy of five marks. It demonstrates relevant knowledge and a sound understanding of the processes of vacuum forming. The answer is logically sequenced and contains most of the correct technical terminology.

The candidate could have improved the answer by adding details of removing the mould from the vacuum-forming machine and by giving details of how it would be trimmed and finished.

Mark scheme

One mark is awarded for each correctly described stage of the vacuum-forming process up to a maximum of six marks.

● Place the mould onto the platen and lower the platen.

● Clamp a sheet of thermoforming polymer onto the vacuum-forming machine.

● Heat the thermoforming polymer until it becomes soft and then raise the platen.

● Switch on the vacuum to suck the softened thermoforming polymer around the mould.

● Remove the heat and allow the thermoforming polymer to cool.

● Remove the mould from the formed thermoforming polymer.

● Trim and finish the edges.

Sample question: Natural, synthetic, blended and mixed fibres, and woven, non-woven and knitted textiles

Fashion and textile designers use a variety of processes and techniques to improve the structural integrity of products.

(a) Give one detailed reason why interfacing would be used on a shaped sweetheart neckline like the one shown in the diagram. [2]

(b) Evaluate the use of piping and boning as methods of improving the structural integrity of textile products. [6]

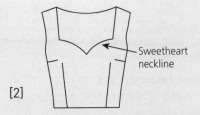

Sweetheart neckline

Figure 22.7 A sweetheart neckline

(a) As the sweetheart neckline is shaped the interfacing will stabilise the fabric [1] and prevent the fabric edge from distorting out of shape. [1]

The purpose of the interfacing has been clearly stated and the reason for its use included – a detailed response worthy of the two available marks.

Candidate response

(a) As the sweetheart neckline is shaped the interfacing will stabilise the fabric and prevent the fabric edge from distorting out of shape.

(b) Most fabrics are flexible and have good draping qualities, however this often means it is difficult to achieve a good shape and structure in some textile products. Piping is a corded component that can be inserted into the seams of textile products. This adds strength to seams and helps stabilise the seam. The cord used in the piping reinforces the shape of the seam.

Boning is inserted into a casing which has been stitched onto fabric wherever support is needed in the intended product. Boning is rigid and will therefore maintain its shape during use. Corsetry, where a rigid supportive shape is needed, is a typical example of where boning is used.

Piping stabilises and strengthens seams whereas boning creates a more rigid structure, however, both improve the structural integrity of textile products.

Boning is more clearly explained and includes reference to its practical use in products. The candidate could have included a similar example for piping, for example, to give more rigidity to the sides of a tote bag.

(b) In responses to evaluate questions evidence of appraisal or making a judgement needs to be clear. There is some evidence of this within this response. This answer would gain full marks within band 2.

The candidate demonstrates good knowledge of both piping and boning when used to improve the structural integrity of textile products and clearly states a reason why these methods would be needed.

Reference to piping is recall of facts and does not explain how piping strengthens and reinforces a seam. Further explanation is needed. For example, the extra fabric used in piping and additional stitching increases the seam's strength. The cord in the piping is often quite thick and would prevent an edge seam from folding over – helps maintain a shape.

Mark scheme

Band 3	A coherent answer demonstrating detailed, relevant knowledge and understanding to evaluate how designers can use piping and boning to improve the structural integrity of textile products. There will be evidence of relevant examples and well-developed substantiated judgements in a response which is logically structured.	5–6
Band 2	Answer has some coherence, demonstrating partial knowledge and understanding to evaluate how designers use piping and boning to improve the structural integrity of textile products. There will be some evidence of mostly relevant examples and partly substantiated judgements in a response which is generally well structured.	3–4
Band 1	Answer demonstrates only basic knowledge and understanding to evaluate how designers use piping and boning to improve the structural integrity of textile products. There will be limited evidence of relevant examples or judgements in a response which demonstrates little structure.	1–2
	Award 0 marks for incorrect or irrelevant answers.	

Glossary

A/B testing: user testing to choose between two different design ideas.

Additive manufacture: computer-controlled manufacture of a 3D object by adding materials together layer by layer.

Algorithms: a logical computer-based procedure for solving a problem.

Alloy: a mixture of two or more metals.

Amplifier: a subsystem to increase the amplitude of an analogue signal.

Analogue sensor: a sensor to measure how big a physical quantity is.

Annealing: a method of heat-treating metal that makes it as soft as possible.

Anthropometrics: the study of the sizes of people in relation to products.

Aramid fibre: a non-flammable heat resistant fibre at least 60 times stronger than nylon.

Assembly lines: a line of equipment/machinery manned by workers who gradually assemble a product as it passes along the line.

Automated production: the use of computer-controlled equipment or machinery in manufacturing.

Back emf: a high voltage spike produced when motors, solenoids or relays are used.

Batch production: making a small number of the same or similar product.

Batch: a limited number in a set timescale.

Bauxite: ore containing aluminium.

Bespoke: made to measure for an individual client.

Bevel gears: a system to transfer the direction of rotation through 90°.

Biodegradable: a material that will degrade/decompose by natural means.

Biomimicry: taking ideas from and mimicking nature.

Blow moulding: a method of shaping a thermoforming polymer by heating and blowing into a dome.

Breadboard: a temporary method of constructing electronic circuits for testing.

CAD/CAM: computer-aided design/computer-aided manufacture.

Cam: a component used with a follower to convert rotary motion to reciprocating motion.

Capacitor: a component to store charge, typically available in ceramic, polyester or electrolytic types.

Carbon footprint: the amount of carbon dioxide released into the atmosphere as a result of human activity.

Carbon neutral: no net release of carbon dioxide into the atmosphere – carbon is offset.

Cellulosic fibres: natural fibres from plant-based sources.

Chassis: the base supporting structure.

Circular economy: a system that aims to minimise waste and extract the maximum value from resources which are kept in use as long as possible, recovered and regenerated into new products instead of thrown away.

Client: the person the designer is working for (this may or may not be the user).

Closed specification: list of criteria stating what must be achieved and how it must be met.

Cloud-based technology: technology that allows designers to share content via the internet.

Coating: an additional outer layer added to a product.

Colour separation: separate colours (cyan, magenta, yellow and black) printed in different combinations to create other colours.

Compensation: payment given to someone as a result of loss.

Compound gear train: more than one stage of gear train working together, to achieve a high velocity ratio.

Compressed boards: manufactured boards made by gluing wood particles together.

Conductive: the ability to transmit heat or electricity.

Context: the settings or surroundings in which the final product will be used.

Continuous flow production: a non-stop production method which produces high volume parts at low cost.

Conversion: the process of cutting a log up into planks.

Counterfeit: an imitation of something, sold with the intent to defraud.

Cracking: the processing of naphtha into monomers.

Cradle-to-cradle production: considering the product's complete life cycle including the reuse or recycling of materials after its initial use has ended.

Cradle-to-grave production: considering the product's complete life cycle until it is disposed of.

Crimp: the waviness in a fibre.

Criteria: specific goals that a product must achieve in order to be successful.

Cross grain: horizontally across the fabric from edge to edge.

Cultural awareness: understanding the differences in attitudes and values between people from other countries or other backgrounds.

Culture: the ideas, customs and social behaviours of a particular people or society.

Current: a measure of the actual electricity flowing, in amps (A).

Deforestation: the large-scale felling of trees which are not replanted.

Deforming: changing the shape of a material by applying force, heat or moisture.

Design collaboration: a number of designers working together on specific design projects.

Design fixation: when a designer limits their creativity by only exploring one avenue of design or relying too heavily on features of existing designs.

Development: the creative process of selecting ideas, elements, materials and manufacturing techniques from initial ideas and using them in new ways to explore and produce newer and better designs or ideas.

Digital sensor: a sensor to detect a yes/no or an on/off situation.

Driven gear: the output gear from a gear train.

Driver: a subsystem used to boost a signal so that it can operate an output device.

Driver gear: the input gear on a gear train.

Dual-in-line (DIL): a standard package style for ICs.

Ductility: the property of a material to be able to be permanently stretched out without cracking.

Eco design: designing sustainable products that will not harm the environment by considering the effects of the technology, processes and materials used.

Ecological deficit: a measure which shows that more natural resources are being used than nature can replace.

Ecosystem: the natural and delicate balance of plant life, soil and animals that live interdependent lives.

Effort: the input force on a lever.

Electroluminescent: materials that provide light when exposed to a current.

Embellishments: surface decoration such as embroidery and beading.

Environmental directive: a law to provide protection for the environment.

Ergonomics: the relationship between people and the products which they use.

Exploded drawing: a 3D drawing method to show how parts in an assembly fit together.

Exploitation: treating someone unfairly in order to benefit from their work; to take advantage of.

Extrusion: a length of polymer with a consistent cross section.

Extrusion (process): drawing material through a shaped former to create the shape required.

Fabric construction: the way a fabric has been made.

Fabric finishes: these are added to fabrics to improve their aesthetics, comfort or function. These finishes can be applied mechanically, chemically or biologically.

Fabric specification: sets out the requirements of the fabrics needed for a product.

Fad product: a product that is highly popular for only a very limited amount of time.

Fast fashion: a recent trend copied from the catwalks, often low quality and low in price.

Feedback: achieving precise control by feeding information from an output back into the input of a control system.

Felling: the process of cutting down trees.

Ferrous metals: metals that contain iron.

Fibre: a fine hair-like structure.

Filament: a very and fine slender thread.

Finite fossil fuels: a limited amount of natural resources that cannot be replaced.

Flowchart: a graphical representation of a program.

Fluidised bath: blowing air through a powder to cause it to behave like a fluid.

Focus group: group of people used to check a product design is on track.

Force: a push, a pull or a twist.

Forest Stewardship Council (FSC): an organisation that promotes environmentally appropriate, socially beneficial and economically viable management of the world's forests.

Fractional distillation: the processing of crude oil into naptha.

Frequency: the number of pulses produced per second, in hertz (Hz).

Fulcrum: the pivot point on a lever.

Generative design: a computer-based iterative design process that generates a number of possibilities that meet certain constraints, including potential designs that would not previously have been thought of.

Geotextiles: textiles associated with soil, construction and drainage.

Global warming: a rise in temperature of the Earth's atmosphere caused by gases and pollution.

Grams per square metre (gsm): the weight of paper and card.

Green timber: timber that has just been felled and contains a lot of moisture.

Haematite: ore containing iron.

Handle: how a fabric feels when handled.

Hardening: a method of heat-treating metal that makes it hard but brittle.

Hardwoods: timber that comes from deciduous trees and is generally harder than softwood.

High-fidelity prototype: a detailed and very accurate prototype, similar to the final product.

Hydrophilic membrane: a solid structure that stops water passing through but at the same time can absorb and diffuse fine water vapour molecules.

Integrated circuit (IC): a miniaturised, highly complex circuit with small pin connections in a single component.

Interactive textiles: fabrics that contain devices or circuits that respond and react with the user.

Interfacing: a bonded fabric, which when fused to a different fabric adds strength.

Isometric drawing: a 3D drawing method, using 30° angles for projections of depth.

Iterative design: a repeated cycle of quickly implementing designs or prototypes, gathering feedback and refining the design.

Jigs: mechanical aids used to manufacture products more efficiently.

Laminated boards: manufactured boards made by layering sheets together.

Landfill site: a rubbish tip, where waste is simply buried.

Latching: a switch which stays on when released.

Lay plan: how templates are laid out on fabric.

Lever: a rigid bar that pivots on a fulcrum.

Life cycle: the stages a product goes through from beginning (extraction of raw materials) to end (disposal).

Life-cycle analysis: an assessment of a product's environmental impact during its entire life.

Light-dependent resistor (LDR): an analogue component to sense light level.

Linear economy: raw materials are used to make a product and waste is thrown away.

Linear motion: movement in a straight line.

Lining: a layer of fabric in the same shape as the outer layer which adds support and conceals the construction processes.

Linkage: a component to direct forces and movement to where they are needed.

Load: the output force from a lever.

Low-fidelity prototype: a quick prototype that gives a basic idea of a product's looks or functions.

Lustre: a gentle shine or soft glow.

Managed forest: a forest where new trees are planted whenever one is cut down.

Manufactured boards: sheets of timber that have been manufactured to give certain properties.

Market pull: a new product is developed in response to a demand in the market or users.

Mass production: large quantities of identical products are manufactured on a production line.

Mechanical advantage: the factor by which a mechanical system increases the force.

Mechanism: a series of parts that work together to control forces and motion.

Member: an individual part in a structure.

Microcontroller: a miniaturised computer, programmed to perform a specific task, and embedded in a product.

Micro-encapsulation: tiny microscopic droplets containing various substances applied to fibres, yarns and materials, including paper and card.

Microfibre: an extremely fine, specially engineered fibre; about 100 times thinner than a human hair.

Micron: one thousandth of a millimetre (0.001 mm) – used to specify the thickness of board.

Milling: cutting grooves and slots into metal.

Modelling: trying out and testing ideas or parts of designs by making scale models.

Momentary: a switch which turns off when released.

Monocoque: structural system where loads are supported through an object's external skin.

Monomer: a molecule that can be bonded to others to form a polymer.

Motion: when an object moves its position over time.

Natural polymers: polymers that are sourced from plants.

Non-ferrous metals: metals that do not contain iron.

Oblique drawing: a basic 3D drawing method, using lines projected at 45°.

Obsolescence: when a product is out of date or no longer useable.

One-off production: only one, unique product is produced; used when making a prototype.

Opacity: lacking transparency or translucence.

Open specification: list of criteria the product must meet but not specifying how it must be achieved.

Operational amplifier (op-amp): a specialised IC amplifier component.

Ore: rock which contains metal.

Oscillating motion: movement back and forth in a circular path.

PAR: planed all round.

PBS: planed both sides.

Perspective drawing: a realistic 3D drawing method.

Phase-changing materials: encapsulated droplets on fibres and materials that change between liquid and solid within a temperature range.

Programmable interface controller (PIC): A microcontroller used in many products.

Polarised: a component which has positive and negative leads and which must be connected the correct way around in a circuit.

Polymer: substance or fibre with a molecular structure made up of smaller units bonded together.

Polymerisation: the blending of different monomers to create a specific polymer.

Potential divider: two resistive components used to provide an analogue input signal for an electronic system.

Preservative: a chemical treatment applied to wood to prevent biological decay.

Press forming: a method of shaping a thermoforming polymer by heating and pressing into a mould.

Primary data: research collected 'first-hand' by yourself.

Primary user: the main user of the product

Primer: the base coat of paint applied straight to the material surface.

Printed circuit board (PCB): a board with a pattern of copper tracks which complete the required circuit when components are soldered on.

Program: a set of instructions which tells the microcontroller what to do.

Protein fibres: natural fibres from animal-based sources.

Prototype: an early model of a product or part of a product to see how something will look or function.

PSE: planed square edge.

Pulp: raw material from trees used to make paper.

Qualitative data: observations and opinions about a product.

Quantitative data: specific measurable data given in numerical form.

Quantum tunnelling composites: materials that can change from conductors to insulators when under pressure.

Rapid prototype: an accurate, finished 3D part produced from a CAD model in a matter of minutes or hours.

Rating: the maximum specified quantity a component is designed to handle.

Raw edges: fabric edges that are not neatened – unfinished.

Ream: pack of 500 sheets.

Reciprocating motion: movement back and forth in a straight line.

Recycled paper: paper made from wood pulp using some re-pulped paper.

Reinforcement: extra material added to increase strength.

Rendering: the method of using colour and shading to represent the nature of a surface.

Resist method: a means of preventing dye or paint from penetrating an area on the fabric. This creates the patterns.

Ribs: thin supports applied at right angles to a sheet to increase its rigidity.

Rigid: inflexible, stiff, will withstand forces without bending.

Rotary motion: movement in a circular path.

Rotational velocity: the number of revolutions per minute (rpm) or per second (rps).

Schematic diagram: a circuit diagram showing the connections between individual components.

Seam allowance: the distance between the raw edge of the fabric and stitching line.

Seasoning: the process of removing moisture from newly-converted planks.

Secondary data: research collected by others.

Self-finished: a material that does not require the application of a finish to protect it or improve its appearance.

Selvedge: the sealed edge of the fabric.

Shot-blasting: using grit, fired at high pressure, to clean a surface by abrasion.

Simple gear train: two spur gears meshed together.

Smelting: the process of extracting metal from ore.

Softwood: timber that comes from coniferous trees and is generally less expensive than hardwoods.

Spur gear: a gear wheel with teeth around its edge.

Stakeholder: a person other than the main user who comes into contact with or has an interest in the product.

Stock forms: the standard shapes and sizes of materials that are commonly available.

Straight grain: indicates the strength of the fabric in line with warp yarns.

Structure: a collection of parts to provide support.

Subroutine: a small sub-program within a larger program.

Subsystem: the interconnected parts of a system.

Surface mount technology (SMT): the industrial method of mounting miniature components onto a PCB using robotic machines.

Sustainability: meeting today's needs without compromising the needs of future generations.

Sustainable design: a design strategy which minimises the environmental impact of manufacturing and using the product.

Synthetic: derived from petrochemicals or manmade.

Synthetic polymers: polymers that are sourced from crude oil.

System: a set of parts which work together to provide functionality to a product.

System diagram: shows the interconnections between subsystems in an electronic system.

Systems thinking: considering a design problem as a whole experience for the user.

Technology push: products developed as a result of new technology.

Tempering: a method of heat-treating metal that reduces brittleness.

Tessellation: an arrangement of shapes closely fitted together in a repeated pattern without gaps or overlapping; used to plan the cutting of shapes using the least amount of material.

Thermistor: an analogue component to sense temperature.

Thermoforming polymer: polymer that can be softened by heating, re-shaped and set over and over again.

Thermosetting polymer: polymer that can only be softened by heating and shaped once.

Throwaway culture/society: refers to the culture of buying affordable fashionable items, wearing or using them for a short period of time before disregarding them or throwing them away; little thought given to the environmental impact.

Tolerance: an allowance included in the seam allowance for inconsistency when assembling a product.

Torque: a turning force.

Triangulation: the method of increasing rigidity by designing a structure to compose of triangles.

Turning: a method of producing cylinders and cones using a centre lathe.

Ultraviolet (UV) light: outside the human visible spectrum at its violet end.

User: the person or group of people a product is designed for.

User-centered design (USD): looking at and checking the needs, wants and requirements of the user at every stage of the design process.

User testing: testing by observing a user interact and use your product for its intended purpose.

Vacuum forming: a method of shaping a thermoforming polymer sheet by heating and sucking around a former.

Velocity ratio: the factor by which a mechanical system reduces the rotational velocity.

Veneer: thin sheet of natural timber used to cover manufactured boards.

Virgin fibre paper: paper made entirely from 'new' wood pulp.

Voltage: the electrical 'pressure' at a point in a circuit, in volts (V).

Voltage gain: the amplification factor of an amplifier subsystem.

Warp: yarns that run along the length of fabric.

Wasting: the process of shaping material by cutting away waste material.

Weave: the pattern woven in the production of fabric.

Weft: yarns that run across the fabric.

Worm drive: a compact gear system which achieves a very high velocity ratio.

Yarn: spun thread used for knitting, weaving, or sewing.